When Law and Medicine Meet: A Cultural View

INTERNATIONAL LIBRARY OF ETHICS, LAW, AND THE NEW MEDICINE

VOLUME 24

The titles published in this series are listed at the end of this volume.

WHEN LAW AND MEDICINE MEET: A CULTURAL VIEW

by

Lola Romanucci-Ross

University of California,
San Diego, CA, U.S.A.

and

Laurence Tancredi

New York University,
NY, U.S.A.

KLUWER ACADEMIC PUBLISHERS

DORDRECHT / BOSTON / LONDON

A C.I.P. Catalogue record for this book is available from the Library of Congress.

ISBN 1-4020-2756-7 (HB)
ISBN 1-4020-2757-5 (e-book)

Published by Kluwer Academic Publishers,
P.O. Box 17, 3300 AA Dordrecht, The Netherlands.

Sold and distributed in North, Central and South America
by Kluwer Academic Publishers,
101 Philip Drive, Norwell, MA 02061, U.S.A.

In all other countries, sold and distributed
by Kluwer Academic Publishers,
P.O. Box 322, 3300 AH Dordrecht, The Netherlands.

Printed on acid-free paper

Printed in the Netherlands.

CONTENTS

PROLOGUE

The yellow taxicab bolted down Park Avenue in Manhattan as though powered by jet engines. Within minutes it passed 72nd, then 71st, 70th, and Hunter College, and at 68th Street, a garbage truck extending into the east side of Park caused the taxi to swerve to the left, where it hit a delivery truck coming from the opposite direction. The taxi folded like an accordion, killing the passenger. On impact, the delivery truck veered sharply to the right out of control on to the pavement, killing a pedestrian who was walking on Park.

It was a busy weekday in October 1966, before passage in New York State of no-fault automobile legislation. School was in session, and at least twenty students, faculty and passersby witnessed the accident. On interviews, each of the observers had different renditions of what had happened. One witness claimed the taxi was going very fast, and when it reached 68th it jumped the curb, killed the pedestrian, then turned sharply inward hitting the delivery truck. The presence of the garbage truck on 68th was ignored. A second witness saw the delivery truck as responsible for the accident. He claimed that the taxi was going at a reasonable speed, but the delivery truck straddled the midline, extending into the east lane sufficiently to cause the contact with the taxi. This, according to the witness, resulted in the taxi projecting to the west and jumping the curb to kill the pedestrian. One witness insisted that the garbage truck caused the accident between the taxicab and delivery truck, and it managed to skip the curb and kill the pedestrian.

The scenarios of what occurred were like a course in combinations and permutations. Each of the witnesses had a different story. Each saw the events from his or her own perspective, distorting reality in many cases to fit pre-conceived ideas of cause and effect. Herein we see the dilemma of determining the nature of a fact, particularly the cause and effect relationships between antecedent events and adverse outcome. Many factors affect how a fact is understood; observer perceptions are shaped by cultural and educational backgrounds, and the context of the event. Ideological viewpoints influence considerably how an individual will concatenate cause and effect relationships. After all, a fact is part of a story, and we know that metaphor, metonym, even myth is used liberally to run the gamut of what connects content with feelings to relay a narrative.

Let's complicate this scene further by transporting it to the year 1995, and adding a criminal action for manslaughter against a party to the accident. Information has been introduced that points to a history of erratic, impulsive behavior on the part of the delivery truck driver, Mr. X. A DNA fingerprint of the participants involved in this tragedy showed that Mr. X had an extensive police record for many skirmishes. He was implicated, but never held responsible, for two prior accidents, and he had been charged with assault when he attacked a man eight months earlier at a hockey game in Brooklyn. On interviewing neighbors, the police learned that Mr. X was known to be "hot headed" and to frequently lose control. At the request of the police and without appropriate legal counsel, he submitted to a Positron Emission Tomography Scan, which revealed that he had abnormal metabolism of his left temporal lobe. This finding was seen by the forensic experts to be consistent with recidivistic dyscontrol, which could explain the sequence of events leading to the accident and death of two people.

At this point, the science seems to complete the narrative; certainly, a "true believer" in biological sciences would view it that way. As with any human endeavor, the mythical and ideologically grounded views of the interpreter strongly influence his conclusions. So, by no means does the science provide a full proof explanation. The facts applicable to the causation of the accident arguably are separate from the condition of any one of the participants. Even if the facts supported the position that Mr. X was out of control, the science associated with both brain imaging and DNA is not unequivocal. Regarding brain imaging, there are many conceptual unresolved issues such as the validity of the laboratory test, the assumption that a particular brain pattern is associated with brain "pathology", and ultimately that such patterns comport and align with aberrant behavior. Similarly, DNA determinations are fraught with the possibilities of error, such as laboratory inaccuracies and confusing the origin of specimens. Neither of these tests is free of interpretative bias, which is highly subjective. Both can, in any one case, produce ambiguous and murky results.

The relationship of cause and effect is also open to scrutiny. Such relationships must be embedded in a preordained theory to be properly interpreted. In most civil cases, for example, cause and effect are established by the showing of a "preponderance of evidence" that "A" caused "B". This constitutes a mere 50% or better showing that a designated cause actually brought about a specified outcome. Other tests with legal significance are that there is a "substantial", "sufficient" or "necessary" relationship between cause and effect. Such tests of causation would likely not fit the average person's view of the cause of an event. In the illustration above, the observers' notions of cause and effect are more strictly identified, most particularly that the cause (reckless driving) was solely responsible for the outcome (accident). This pattern fits the more classic "but-for" relationship, which is that "but-for" the taxicab's recklessness (or the garbage trucks actions) the accident would not have occurred. The causative relationship to the average person requires proximate and direct connections between the act and the event.

But rarely does such a strong relationship exist, except in the context of a legal test that strictly delimits the narrative. Most often, because of the loose nature of stories surrounding facts, the causative factor is seen as a cluster impact, which is to say many factors ("A", "B", "C" etc.) that converge to bring about a particular outcome. At this level of multi-factorial involvement, elements of culture and ideology embed the dynamics between causes and effect to complicate and preclude a pristine determination of what factors are responsible for the effect. In the illustration above, the weather and time of day of the accident are considerations that could powerfully influence how we interpret the facts of the accident.

When Law and Medicine Meet: A Cultural View challenges many of the traditional presuppositions about the power and validity of scientific evidence, particularly bio-medical evidence. This book, written from the perspective of a cultural anthropologist, examines the nature of scientific facts, the impact of culture, myth, and ideology in which these facts are embedded on their interpretations, and the meaning given to facts in the context of stories in which they appear. Chapters 5 through 8 are scientific case studies of three broad technologies, those involved with the assessment and disposition of sexual predators, brain imaging in forensic controversies, and DNA fingerprinting. These studies involve biomedical science along a wide spectrum. On one pole is the "soft" data associated with information on sexual predators where heavily value-laden notions such as "normality" and "deviance" predominate in defining the terms for evaluation and decision-making. On the other is the "hard" data of DNA fingerprinting, which can be highly accurate and predictable, relying on sharply defined terms and strict biochemical methodology. And in between is the scientific data associated with brain imaging, where the various technologies involved, such as PET, MRI, CEEG and SQUID, offer the externality of precise, predictable method, but are fraught with difficulties in interpretation which can dramatically skew the results.

The last two chapters integrate the broad principles from the previous chapters and apply these to a field case study, the O.J. Simpson case. This legal case, perhaps the most publicized trial of the past 50 years, offers a ready made opportunity to scrutinize the validity of factual determinations from everyday life to the most sophisticated of biological sciences. During the trial, both in and out of the courtroom, the arguments for and against O.J. Simpson took place in an environment of massive television, radio, and newspaper commentaries pitted against divergent ideologies, a full range of cultural themes from infidelity to abuse, and conflicting racial attitudes. The American public was riveted to the television as the drama unfolded, and their opinions seemed shaped by frequently unproven accusations and claims. Both the events in the courtroom and those in the public forum illustrate strikingly the complexity incumbent in determining the reliability, validity, and relevancy of facts, be they the actions of people in everyday life, or the product of sophisticated scientific inquiry, such as brain imaging and DNA fingerprinting. Even in this modern scientific age, myth, folklore, bias and ideology can influence the way science is presented and interpreted to arrive at important social and judicial decisions.

CHAPTER 1

PRIMITIVE YET CONTEMPORARY: MATRICES OF MEANING

Anthropology embodies the concept of culture, a concept that is ever expanding as it includes all the material and intellectual products of human endeavor. It encompasses systems of thought and of investigation, systems which in turn, influence and shape each other. Thus, thinking about "scientific evidence" and the law convinced us that our approach had to be an anthropological one. In pursuit of understanding cultural differences and biological determinants of human behavior (the so-called Nature/Nurture controversy), anthropologists have constructed a discipline which has become a mosaic of sub-disciplines, and the resulting specialization has led to considerable progress. Medical Anthropology is now a well-established sub-discipline, and will provide a framework for our inquiry.

From describing and analyzing other cultural systems of curing and healing, anthropologists went on to study the dynamics of patient choice in a situation where two or more cultural systems interact (Romanucci-Ross 1969). This led to cultural self-reflection and to studies of various aspects of our own medical system (e.g. Katz 1981, 1999; Lock and Gordon; Kleinman; Trostle), as well as to our medical system within the new context of globalization (Sinha). Medical discourse also has been analyzed in anthropological studies that deal with the relationship between linguistics and power (Kuipers 1989; Foucault 1972, 1973); ecological adaptation (McElroy and Townsend); medicinal plants (Moerman; Etkin); nutrition (Bogin); transcultural psychiatry (Devereux; Tancredi; Kleinman); and some of these fields have now become established as separate sub-disciplines; (for an overview see Romanucci-Ross, Moerman and Tancredi 1997). Some research has been done by anthropologists on systems of law in simpler cultures (Melton; Nader). These and similar studies have been important, for, in such cultures, legal systems (rules of conduct) rest heavily on ethical codes and sometimes "over express" these codes.

In simpler societies, medicine and law (the latter being acceptable conduct and penalties for non-observance) are not distinguishable in praxis. The diagnoses may be brief, require inquiry and time, but eventually the cause of an illness or disability is unraveled and shown to be an infraction of expected behavior, conduct that was prescribed or proscribed toward a person, a place, a spirit, or a species. The latter could be disastrous even for other family members for whom that species is a totemic ancestor. For example, in the Admiralty Islands of Melanesia (South Pacific), and for one linguistic group (the Sori), if one's totem were the osprey or manuay one should not kill or even walk under the shadow of that bird. An offense would guarantee that a child in the totemic group might die, or would never speak if it lived (Romanucci-Ross 1983). One important facet of this exercise is that the judgment is in the diagnosis; in the illness rests the punishment, and the rectification is in the cure. The interplay of law and medical events in such instances of illness and cure reinforces the interdependency of systems in maintaining the social and cultural fabric of meaning.

Such considerations led to our interest in exploring the characteristics of the law and of medical science in our complex culture as we seek to determine the extent to which one system may shape and define the other, and to examine how events external to one system or the other could impact on both systems. This approach has the added value of extending an important tenet of anthropological research: that an understanding of simpler "exotic" cultures may help us to better appreciate the interconnectedness and deep structural functions of our own institutions.

In each profession, the "card carrying" professional has reached the terminus of a long process in which one learns a new language, new ways of relating to self and other (whom to include and whom to exclude), new rituals, and even the manner of dress. (The New York Times in a recent issue had a full-page narrative with photographs and sketches on how lawyers should dress). Most importantly, one learns to adopt new ways of thinking, together with a (perhaps unfortunate) simultaneous contempt for other ways of thinking. Law schools promise they will teach students to "think like a lawyer"; medical schools will teach students to think as physicians or clinicians, or both. We hoped to learn how these very different thought systems relate to deeply structured and embedded epistemologies (theories of knowledge, e.g. in each field there are rules to be learned for which "logical" connections are permissible and others which are not). It may seem trivial and expected that two individuals can communicate clearly. However, communication relies on common meanings and symbols, knowledge of the rules for their use in a single situation, and for conjoining rules in two or more situations. In our systems of law, one combines cognitive and normative expectations in operations, and so, reasoning in this sense is self-referential. Of course, the "problem" that calls for a legal judgment comes from outside, from the broader society, but the rules for solutions are internal to the system of law. Medical science, on the other hand, must go outside of its own system for evolving solutions and new concepts (e.g. bioengineering, molecular biology, space exploration, military exigencies or even the pharmaceutical industry--all are currently influential in suggesting new pathways for medical research). In a report from the workshop on scientific fraud and misconduct for the American Association for the Advancement of Science, it was noted that, for lawyers "the tendency was to take good data at face value" while scientists "are frequently suspicious of data that are 'too good'"; in addition, "lawyers have argued that the absence of primary data strengthens the case of a scientist whose work is impugned, but scientists take missing data as a significant indicator of misconduct" (Woolf p. 71).

Legal reasoning does introduce social constraints, for example the law determines who can and cannot practice medicine, and how medicine is practiced. Further, the law affects the fiscal environment in which medicine is practiced. Also, because ethical problems cannot be resolved in an objective manner, the law must define to some extent what ethical boundaries should be. For example, the court recognized a "liberty interest" in the acceptance or rejection of medical intervention, including the "critical element" of "the right to define one's own concept of existence" (Quinlan).

Can two systems premised on vastly differing assumptions, corollaries and combinatorial rules ever resolve problems they are conjointly asked to address? And if they can, what are the rules that apply to definitions, operational methods and standards for mutually acceptable solutions? Certainly, problems involving the science of medicine and a body of laws are being "resolved" (i.e. settled) every day with every deliberation and every decision reached by judges in our courts. Often, these are settled on the basis of what the scientific evidence is considered to "prove". But outcomes are quite often unsatisfactory to one side or the other, sometimes to both sides of the litigation. It would appear, then, that the process of the making a proper decision and the ingredients that go into that decision ought to be scrutinized in a manner that might lead to a different approaches.

In striving for cohesiveness laws are woven into a mosaic, based on precedent from earlier decisions and the occasional legal challenge to these precedents. From whence then, does substantive change come? Just as in all other systems, change comes from "noise" outside the system, in this case from social, cultural and political change. These movements reconfigure the emotional tone and the structure of "knowing" in a culture, and because of this, changes resonate within other systems in the culture.

"THE NOISE"

In communication theories about language and other symbolic systems, one speaks of "noise in the channel" when there is interference in the encoding and decoding process; such interference introduces the possibility for error. Errors may be caused by mistakes in the codes, by the transmitter or the receiver of codes, or by disturbance in the channel: all these possible causes are usually considered as a set and referred to as "noise" (Miller p. 7). Whatever the specific cause, the effect is the same: the message is distorted and reinterpreted, creating something new in personal or group orientation, and perhaps in behavior.

Interestingly, "noise" had the same meaning to the people in a case of dramatic culture change in a relatively isolated "primitive" society. We refer to the Manus people who live in the Admiralty Islands of Melanesia in the South Seas (1), and who in the period from 1946 to 1952 experienced a type of millenarian movement called a "cargo cult". There had been similar nativistic movements in Africa, in the Americas, in Polynesia, Asia and Indonesia (Lanternari). But those in Melanesia were more akin to messianic movements; that is, the cargo cult featured the belief preached by the local messiah that the dead ancestors would return soon, bringing with them new governance, new laws, good health and well-being, and all the riches (the "cargo") that the white man possessed. Such a messianic figure was Paliau Moloat. Because of a coincidence between planned fieldwork and a period of dramatic cultural change, that period and this leader's activities, his evolving concepts and related events were carefully documented and analyzed (T. Schwartz). Also, there was a longer period of study recorded for this culture area (2). Further, these same Manus people studied in 1928 (Fortune; Mead) had what was described as a long history of traditional rules of kinship, warfare, and conduct, as well as rules about illness, health and healing which related to "laws" about conduct. Sanctions for legal transgressions were carried out by the ghosts of one's lineage or by the malevolent ghosts of another lineage, who exercised control over sexual matters and exchanges prescribed for marriage, or exchanges of goods in other situations. These people had been described by Mead and Fortune as "puritanical" in the strict sense, as understood by Americans in the context of their own cultural history. Malicious ghosts would carry out the punishment by inflicting illness or death. The system appears to have satisfied everyone's notion of justice, until certain chance events occurred that brought this society into a metastable state, and then to a period of confusion and hard choices.

Although the Islands were sighted in the sixteenth century, and "discovered" in the eighteenth century, ships did not stop there until the 1800's (Jacobs). In 1884, certain parts of the Admiralty Islands were "pacified" by Germans who began plantations and established a sort of administrative government. The Germans did not seem to have much of a cultural impact, though they left some interesting writings on what they found there (Nevermann). One stinging memory provided by the Sori of the North Coast of Manus tells of their learning, to their amazement, that Europeans thought they had bought an island when they offered a gift of beads. They wondered to Lola Romanucci-Ross, Who would ever think that any group would sell their only home on this earth for beads? They had heard from missionaries that the Germans wrote in their books that they had purchased certain parts of the Islands for beads and trinkets, and they asked this investigator to correct that (Romanucci-Ross 1983). After 1914 the administration was taken over by Australia, by Japan early in World War II and then, by the end of the war, it was back in the hands of Australia. Missionary activity began at the turn of the century, and during the 1930's there was a flurry of "conversions" by Catholics, Lutherans, Seventh Day Adventists and Evangelicals. But it was World War II and the presence of the American military (Manus was a staging area for battles in the Pacific) that made an indelible impression. Who had ever seen such ships, airplanes, so much clothing, cigarettes and voices (coming out of boxes, not people), all those lights, and everything so incomprehensible on such a scale? Furthermore, Whites had so many laws and rules about behavior, about religion, about how white people take care of their sick; there must be a connection that could be explained in terms of cause and effect. White people must know the secret of how to get things out of their Dead Ancestors, and how to foil the plans of the malicious ghosts of other lineages. And so the white people were observed, and many attempts made to emulate their behavior. Only in illness and curing was there a resistance to total belief in the success formula, and because illness and curing were linked to rules of conduct, this particular aspect of the foreign culture became the center of much confusion (Schwartz, Lola R. a.k.a. Romanucci-Ross).

The Manus people of the Admiralty Islands experienced the sights and sounds of over one million American service men and women for the duration of the war. They saw Blacks working alongside Whites in similar uniforms! Americans were described, in episodes they recounted, as friendly, helpful (saved sick babies), generous (gave them all Coca-Cola drinks). It was amazing to hear of all these encounters in exquisite detail. (They took no field notes, of course, but their motivations and aspirations, coupled with an overwhelming passion to learn the secrets of White culture "success", made the memories indelible for many years). In an attempt to understand the logical connections of behavior and outcomes, they looked for explanation within their traditional culture. Mediums, in curing, would call forth the spirits of the dead. One could go into trance and learn what to do, because appeasing the angry ghosts was the only way to have things come out right. Paliau was a leader in this early cult activity, urging his followers to get rid of the past, by destroying traditional things and ways, and waiting for the good things to come with the cargo. They did this with some regularity, waiting on a WWII American tarmac for the "cargo" plane sent by the ancestors. When this did not happen, the explanation was that they had done something wrong procedurally and this strengthened the belief, rather than causing it to abate.

Paliau, although a leader in the "enthusiasm" phase that began it all, then became determined to transform this ferment into a political movement, so that there would be real changes in their lives in the real present. He started to attack cargo cult fantasies even as he began to organize communal agricultural and other trade activities, forcing concessions from the Australian government; he spent much energy conceiving and implementing new rules of conduct. This was a first step into cultural integration, and needless to say, a large step toward the acceptance of Australian concepts of law and of political behavior. They began to have polling booths somewhat later, but the ideas began to be understandable, even though, in the mid-sixties, it had to be explained to several men on Sori Island that "fish" on the ballot did not mean they were voting for fish, but rather for a candidate that was indicated by a picture of the fish, and that by voting for fish they would not vote for coconut (the other candidate). It made sense when one of the men said, "I think we need both fish and coconuts and I think we need both of these men to work politically for us. It is not right to make one a winner and one a loser if they both want to work for us" (Romanucci-Ross 1983). As Americans, we might have learned something from this, if we can believe that others in the world have something to teach us. In this instance, we might learn that zero-sum games in power politics do not bring out the best in either voters or candidates.

Admittedly, we have no such striking examples in our society, but we study simpler societies because they heighten awareness of parallels in our own. Examples of "noise in the channel" can be found in almost any era: the Civil Rights movement of the sixties, for example, had a substantive impact on laws (e.g. relating to voting rights), but the "noise" from these events radiated to other opportunity spaces in our society. The Civil Rights concept, when applied to women and other minorities, the elderly (Tancredi and Romanucci-Ross 1997), and notably to the rights of the mentally ill and disabled, caused substantive and enduring changes in legislation in these areas.

MESSAGES IN MANIFOLDS

"Noise", with its attendant events, revelations and insights, puts into sharp relief the processes featured in culture change. The Manus had a functioning social structure, which included natural phenomena: plants and animals as totemic "starters" for genealogical lineages, ghosts of the Dead and many additional "other than human" agents. All these played an important role in decision-making, in explanation of cause and effect in health and illness, as well as in behavior. The structure served satisfactorily until the element of "chance", in this case World War II, provided a major stimulus for system perturbation. As Manus became an operational base for American forces, what the native people saw and heard was overwhelming, as indicated above. As a consequence, their goal became to enhance their knowledge about how to get what they longed for, what the Americans and other Whites had: the material means for a better life. But they were guided only by their own logic when confronted with assimilating this new information. The thought system they knew proved insufficient to make sense of what was observed. Therefore, séances and trance behavior, traditionally the province of mediums who communicated with the Dead to receive news or learn how to end an illness, now became the channel for communication of local political village figures. Also, cleaning and rearranging cemeteries was undertaken; this could be seen as searching for a new way of thinking with new rules of cause and effect. It also exemplified a new and forceful phase of a previously hesitant and unenthusiastic acculturation process. It now seemed to them that the "prize", receiving the cargo, was at hand, and the "price" was logical and right: do what the American military did, simply clear the land for landing fields and wait. Furthermore, demonstrate to the dead ancestors that you are equal to the task, and abandon all the detritus of your earlier failed methods. There is much in the cargo cult phenomenon that is instructive, if one can get past the ambush laid by quaint lay writings by authors who dwell on the curious "primitive" aspect of building an airstrip and making fires to light the way, while the hopefuls wait for planes that do not, and will not, arrive (see Huber p. 56-58) as an example of such exemplifying). Given the complexity, an interpreter not well versed in the area should be admonished not to focus on a minor "busywork" aspect of profound cultural adjustment to modernity. For some strange reason not apparent to us, Huber suggests that waiting for the cargo to arrive in New Guinea is comparable to trial lawyers putting junk scientists on the stand. This comparison does not stand. (At least not to anyone who has done anthropological field research in cargo cult country.) Therefore, let us examine more closely how the "noise" and "cargo" concepts might more appropriately be applied to our purposes.

ADJUSTING THE PARALLAX ERROR

Legal issues in medicine, or medical science issues in the law, are indeed defined in larger frameworks that are political, social, and economic, all of which are culturally driven. On one level these issues explore what we, as a society, consider to be the good life, the better life, or the good death, and issues of control in the context of health care and therapeutics. With this in mind we note current "flash points" of medico-legal interaction e.g. patients' rights legislation, civil rights of the mentally ill, notions of privacy protection, women's health (with all the new interpretations introduced by the feminist movement). Another movement, led by those who would protect the environment, has also brought to the fore health care issues and legal rights to health protection in the work place, and where workers must live because of income and proximity to the workplace (Harr). In these contexts we wonder, as did the Manus New Guinea natives, why the good things that are supposed to be available to all, keep eluding us. For reply, we look to our intermediaries who have been endowed with Authority: doctors, medical scientists, lawyers and judges.

In American society, professionals in both law and medicine have a certain status that is similar in some respects although, in general, physicians appear to be more esteemed than lawyers. In our medical education experiences we have found that medical students, like the general public, have a low tolerance for ambiguity, as well as for uncertainty in complexity. Never mind that Hippocrates, in the First Aphorism, is reputed to have said that life is short, the art of Medicine is long, judgment difficult, the trial of therapy precarious, the crisis grievous and the outcome uncertain. On the other hand, lawyers pounce on and thrive upon ambiguity and complexity. The general public often views the lawyer as distorting ordinary language and torturing meanings, bickering over what they call "rules of evidence" as one side tries to get evidence into the case while the other side tries to keep it out. Therefore, it is assumed that unlike the physician, the lawyer is not seeking the truth but rather works only to benefit his client. In this latter assumption, the public view is correct: the duty of the advocate is to present the case for his client in the best possible way, within the limits of legally acceptable strategies. The basic premise is that even the guilty have the right to a fair trial and due process. It would be helpful if it could be better understood that this very process - the stances taken by litigants as advocates - is a necessary step toward arriving at a "truth". As for the truth, this is to be determined by the "trier of fact", jury or judge. Only these have the formal duty to consider the judgment.

Since Daubert (q.v.) federal judges have been informed that they must weigh and assess expert testimony to be assured that it satisfies the criteria of what is called "good science", as opposed to what some have called "junk science". We maintain that science, when it fits the definition intended for the activity that warranted the label, is practiced by those who know how to identify and apply appropriate and reliable methods to gather data, how to provide a logical framework for interpreting the data, and how to apply acceptable assumptions that relate the data to pertinent theory.

ON "JUNK" THINKING

"Junk science", as discussed in "Galileo's Revenge" (Huber 1995), is exemplified in court cases in which diagnoses or cause-effect phenomena were so far fetched that no one in our times, and even in the times the cases were being tried, would have disagreed with the author on applying the "junk" label. Although Huber cited the large sums that juries awarded as evidence that juries were convinced by the arguments, as in many current cases, the size of these awards express the compassion of the jurors, not the weight of the evidence. Many such awards were subsequently greatly reduced, and many cases were overturned on appeal (both then and now). What interests us here, however, is not an analysis of jury behavior, but rather the view of the author. His focus leaves no room for the needed recognition that even "good science" has its "junk" aspects, which should be extruded by its practitioners. This very process of correcting, and the methods by which it is accomplished, are fundamental to the scientific method. However, since the correction phases are neither predictable nor "scheduled" it does not seem wise to assert with conviction that science can be reliably labeled as either "good" or "junk".

From a cultural standpoint, perhaps the most interesting aspect of "Galileo's Revenge" (Huber 1995) is the timing and the context of its publication. The new Republican Congressmen among the 1992 "freshmen" members wanted a vote on their "Common Sense Legal Reform Act", which included a number of measures to strengthen defenses against tort cases (lawsuits seeking damages). Businesses and insurers wanted this legislation passed, whereas plaintiffs' lawyers were opposed. Business lobbies had long wanted to see many of the features of this Reform Act in place; for example, "loser pays" would prevent a number of "frivolous lawsuits". But "loser pays" would at the same time discourage many ordinary citizens from filing justifiable lawsuits ("frivolous" is always in the eye of the beholder). This Reform Act also had a provision for "honesty in evidence" which would bar admission of expert testimony unless it was based on "scientifically valid reasoning". (Some felt that the courts were not qualified to make such a decision). In addition, defendants (e.g. insurers or industry) would be liable for punitive damages only if it could be proved that there was clear and convincing evidence of malice. Also, there would be an end to joint liability so that, for example, the seller would not be responsible for defective products (New York Times, 1995). Those attempts at reform were brought to a standstill at that time, but they have now reemerged during the George W. Bush administration.

Huber's book constituted an Apologia for the Reform Act movement, and therefore it is strikes the reader as a series of agonistic displays against plaintiffs and against "Calabresians", as he calls those who look for deep pockets to sue; in his view these people want to become involved in social engineering (Calabresi). In our view, Huber's book was not intended to be a scholarly work on the nature of science, junk or genuine. Perhaps for this reason, the book has assumed a greater importance in arousing much controversy and focusing attention on the court as arena for social change through a dialectic between "science" and the legal establishment.

CHANCE ENCOUNTERS

We learn in law school, that law is a hermeneutic science, with meanings to be found in continuous interpretations, and from "unfoldings" of negotiated understandings in past cases. The power behind such exercises (political, social or cultural) can significantly shape the law, and the law itself creates and perturbs power balances. "Informed consent", for example, that cornerstone of patients' rights (as well as a protective guideline for the physician) had evolved from the tort law approach, the goal of which was to avoid injury and battery (unconsented touching) to the patient. The legal test for how much information a doctor needed to disclose to the patient went from the physician disclosure standard to the "reasonable man" standard, to the subjective standard (what a particular patient would want to know). Thus, informed consent currently asks no less of a physician than that he know and understand the particular patient, so that he can comprehend what that patient would want to know about outcomes of specific interventions. These changes were introduced by a series of landmark court decisions (3).

Another dramatic instancing of the resonance quality of culture change from one cultural sector to another can be seen in the progression of civil rights for the mentally ill. Public interest in science had, by the mid-twentieth century, established that when there were differences of opinion, "science" should intervene and prevail. It seemed logical to extend this dictum to the courts. At this juncture in American history there were a number of major social, cultural, and scientific movements that urged and facilitated change in the treatment of mental patients. Public mental hospitals were in a disheveled state on many levels. Political liberals criticized the administration of mental hospitals for lack of proper regulations, patient repression and abuse. Political conservatives wanted to abolish these hospitals because they were too expensive and money could be diverted for other purposes (Grant). It should be noted that such harmonious agreement for "shut down" across the political divide, would never had occurred if ordinary hospitals were at issue and subject to the same criticisms. What rendered the mental hospital so vulnerable is this: the definition of mental illness is, and always has been, negotiable (Romanucci-Ross 1996). And this is so without even considering the legal definition of insanity, which concerns one's ability to appreciate the nature of one's own conduct in reference to specific acts. The 1940's and 1950's ushered in the "deinstitutionalization movement", a consequence of political negotiations, amplified by the appearance of the first effective medications for treatment of psychoses in the late 1950's. Reserpine was an early psychotropic drug; chlorpromazine also became available at that time, as well as the phenothiazines.

The advent of medications, together with political pressures from various groups, and the incipient Civil Rights Movement, led many investigators to categorize the mentally ill along with minority racial and ethnic groups in need of understanding and rectification of past injustices. This led to Community Centers legislation aimed at solving the problems created by the rapidly increasing number of released patients brought about by deinstitutionalization. These movements and their accompanying entitlements led us to the 1990's being declared "the decade of the brain", with high hopes that the neuropathology and pharmacology of the brain would become better understood (Tancredi). Important for our discussion, all of these factors had an impact on changing legislation for the mentally ill, legislation that included their rights as citizens, their rights to privacy in view of the "holding" laws, and the issue of administration of medications.

How do medications affect the patient's right to privacy (4)? In one case in 1981, the court established that a patient had the right to refuse treatment on right to privacy grounds: use of neuroleptic drugs was determined to be an invasion of privacy. In another case, the court established that a patient has a right to treatment once confined (i.e. if one is confined because of an illness, even mental, then the confined person cannot be denied treatment). In the Karen Ann Quinlan matter (Quinlan), the court recognized a "liberty interest". Karen Ann was maintained on a respirator in an irreversible coma with massive brain damage, without hope of return to any level of normalcy. Her father requested removal of the respirator, pitting the family (and family resources) against the courts for years. In "Quinlan" the New Jersey Supreme Court authorized the patient's guardian to order removal of the respirator, provided the patient herself would have chosen that course, thus applying the "right to refuse treatment". The Saikewicz case provided another twist on this quandary as the court pondered what a "competent" person might decide, knowing he is incompetent. The presumption here would be that a person is not competent and is somehow aware of this on another level of conscious knowing. In the case of Richard Roe III the court again returned to that question and concluded that Richard's father could not be a suitable substitute for such a decision; the father, in this particular case, was not a disinterested party, since he argued in terms of effects decisions would have on him and his wife. These and other cases are cited to illustrate the change in emphases from simple and straightforward constitutional rights (refused or denied), to very sensitive, indeed almost existential, views of a person's rights, including the right to one's own bizarre thoughts (4).

TIDES OF TIME

The arts, the theatre and literature of America (and Europe) during "the sixties" were redolent with themes of freedom from repression from any and every source, including freedom from political realities, as well as from personal angst. Expatiating out to more popular writers from the views of such existentialists such as Kierkegaard, Sartre, Marcel, Heidigger, etc. the new ideologies of how the self relates to others and to culture found, in this era, ready audiences in authors, journalists and their readers. All of this had a significant influence on decisions, both administrative and legal, and especially on changing lifestyles. In the sphere of mental health, there were books with titles such as *The Snake Pit* (also made into a film), *The Myth of Mental Illness* (Szaz), *Every Other Bed* (Gorman), and *The Mind that Found Itself* (Beers). *The Doors of Perception* (Huxley), *Magic Mushroom* (Leary) and *The Divided Self* (Laing) provided new vistas of states of consciousness, induced either by chemicals or by a recognition of social and cultural bondage of the self, without prior knowledge or consent. *One Flew over the Cuckoo's Nest* (Keesing) poignantly portrayed cultural misunderstanding in the assignation of madness, as well as the minor tyrannies within the structure of medical authority in a mental hospital. It would be difficult to overemphasize the effect that these messages, received as theatre or literature, had on shared understandings that found their way into legal expression of constitutional meanings. In the arena in which lawyers relied on expert witnesses, it led the courts to turn away from the notion of contract, for it was generally felt that a society had other values in which its "contractual relationships" should be vested. Some insisted that beyond contract lay the notion that liability should fall heaviest on the proximate cause agent who could have prevented the injury at the lowest cost, therefore, should be required to pay (Huber p. 22).

But things did not rest there; the pendulum was due to swing back. It now appeared that although problems with the mentally ill had been somewhat ameliorated, in many ways, the basic problem of getting medical care to those who needed it and social work services to others, were not being solved. Jails and the streets had large numbers of homeless persons who formerly had been hospitalized for mental illness. A counter movement then started aimed at faulting "the mental health lobby" which included civil rights activists, along with groups that were composed of families of the mentally ill (Isaac and Armat). This movement has had some recent successes in 2000 and 2001, at least in California. Provisioned in Assembly Bill 1421, judges can compel a patient to accept professional help; AB 1422 would create a new health commission composed of various professionals in a community; AB 1424 makes it possible for family members to see to it that a mental patient gets the help and medications he or she needs, even if the patient is unwilling (this bill was introduced by H. Tomson, a former psychiatric nurse now in the legislature). It is remarkable that these bills exist in California, where the legal rights of mental patients are still the agenda for "attentions".

In another culture, deinstitutionalization also ran its course over approximately the same period of time. In Italy, however, it became a more unified, celebrated cause, particularly in the North, where it was more politicized. The aftermath of this experiment which failed was studied and commented upon by many over a period of several years. Dr. Basaglia, founder of what he called the "new psychiatry", claimed that madness was the result of the "psychiatrization of poverty". Basaglia (1980) insisted that such poverty, neglected by the State, was nothing but institutionalized violence. His goal was to expose the power structures both outside and within the mental hospitals. To this end, he and his group were instrumental in the passage of Law 180, which attacked the concept of dangerousness that had allowed patients to be institutionalized in the past. This bold frontal attack on the social structure of a nation through its laws engendered many critiques (both positive and negative), numerous articles and books, and studies of what was to be found in the wake of this movement. In this instance, the moment of great cultural error occurred when psychiatric testimony transcended the evidence and worked its way into substantive issues of law (Romanucci-Ross, 1991, pp. 144-146).

What can be said about the nature and the characteristics of "culture" depends on where you choose to enter the cultural system in order to formulate a definition. No culture is a monolith, and the relationship between the general culture of a group, its subcultures and other cultural enclaves, cannot be described as though it were an organizational chart. It can be likened, rather, to linked sets of circuital fields. Neither the general culture nor the groups it "contains" are closed systems. If scientific medicine and the law are to be understood within the parameters of the culture in which they are practiced, consideration must be accompanied by the recognition that scientific medicine and law, when in motion, also introduce changes in the general culture. In what follows, we attempt to describe scientific medicine and law as systems of thought and action, and to exemplify how they are represented in the courts. We use the case study approach as well, beginning with a case that illustrates "the science of commitment" and "therapeutic jurisprudence" as instanced in changing laws and the important analytical questions that emerge from it. This is followed by Brain Imaging as evidence, and questions of test interpretation. What can be inferred from such imaging? Does the image actually represent a physiological condition related to a specific brain function, or is it one of several possible artifacts or other secondary phenomena? Our third case will consider "DNA fingerprinting", as an example of an approach which is considered specific and accurate, yet has a variety of difficulties that can affect test results. Our main focus will be on the influence of cultural factors, including beliefs and belief systems, ideology passing as knowledge, and second realities posing as the Real. Our goal is to analyze the logics of discovery, while taking note of another element that always gets a crack at influencing outcomes..."chance".

NOTES

1. The Admiralty Islands, a small archipelago situated 150 miles North of New Guinea at about 2 degrees latitude and approximately the same distance from the Bismarck Archipelago, includes Manus. In the period 1961-1963, referred to in several references above, the population had reached 20,000.
2. Anthropological field research was begun by Reo Fortune and Margaret Mead in 1928. In 1953-1954, Theodore Schwartz and Margaret Mead continued the research. From 1963 to 1967, the research team consisted of Theodore Schwartz, Lola Romanucci-Ross and Margaret Mead; they extended the research both geographically, (by including the entire Admiralty Islands Archipelago) and in the types of materials to be investigated and analyzed. The 1963-1967 study was financed by the National Institutes of Mental Health and sponsored by the American Museum of Natural History.
3. In general references see: *Nathenson v. Klein*; *Canterbury v. Spence*, and the Mcpherson case.
4. In general references see: *Rogers v. Okin*; *Rennie v. Klein*; *Donaldson v. O' Connor*; *Reise v. St, Mary's*; *Superintendent of Belchertown State School v. Saikewicz*; *In re Richard Roe III*; *Wickline v. State of California*.

CHAPTER 2

THE SHAPE OF BELIEVING

An impressive body of literature exists among the ethnographies of cultures in the world on the many ways man has devised to determine the nature of "the real" and "the true". If we accept Ludwig Fleck's definition of a "fact" as a stylized signal of resistance in thinking (1) then we must accept that a "fact" is conditioned in time and in space (or place) and that it is to some extent observer-dependent. By this we mean that context determines a fact, and that observer insight plays a role as well. When contexts change, our definition of a fact may need to be revised, or, its factual status might even become more securely grounded.

What concerns us here, however, is not a comparative study of various schools of thought about the nature of the factual. Rather, our concern is with the definition and role of a fact in science (especially biomedical science) and in the law. In particular, we note and analyze the matching and mismatching of logics of two worlds of discourse in an arena (the court) when adversarial positions are in play.

Much of the discourse in this arena begins with causality, and concerns arguments in the simplest of cases on the nature of the event, the nature of its cause, and the nature of the circumstances. The advocate with an ideological agenda will choose the "cause" to fit the desired outcome, for, as indicated elsewhere, belief in causality is conditioned by the ideologue's program for "social action".

We can think of belief as a culturally inflected personal guide to our view of the world and "the real". Every aspect of one culture's belief systems is not shared by all members of the culture; each member has a personal version of whatever beliefs that individual has selected from the cultural pool of ideology options. Some views and beliefs are shared by groups within that culture, forming a schemata of cultural sharing. Some elements will be shared by all cultural members, some by most, some by a few. An ideology is an overall condition of learning and of knowing, which is not fragmented into "religious", "political", "social" or even "scientific" beliefs. A belief system encompasses attitudes in all these aspects of a personal field and more often than not contains inconsistencies or is incomplete; but believers live with these inconsistencies and accentuate differences with other belief systems. Belief systems are part of any field of "knowing" as we shall have occasion to explore here; an individual may not even be aware of being a "believer".

Some who write about legal affairs assert and emphasize the "soft criteria" nature of the social sciences in contrast to the "hard criteria" of the natural sciences (2). Not a novel approach, it has been put forth by many others for quite a long time (as did William James in 1909 and 1912). In relatively recent times, this dichotomy has been refuted by a number of scholars (e.g. Nagel, Feyerabend, Broad and Wade, Kuhn), although others persist in speaking and writing about "intuitive" ways of knowing, about description--thick or thin, and about "interpretation". They would maintain that it is the use of such criteria that constitutes the fundamental issue in examining the way the law views events as opposed to the scientific view. We ask, however: are judgments in the natural sciences really different from those in the social sciences? Are "data" the only foundation of the natural sciences? Must data always and only be obtained through an essentially mechanistic method that is experientially based, relying heavily on measurement? In an investigative spectrum of "scientific precision", should "soft criteria" be positioned at one extreme and distinguished from "hard criteria" at the other? Even if we treat a "scientific fact" as a basic unit derived from observation, analyses and interpretation, we must consider the potential role of beliefs, concepts and ideas and how they relate to any scientific fact and the legal understanding of it.

Scientific endeavor was characterized by Frank as having different emphases assigned to observation, to logical reasoning, or to creative imagination (3). This approach makes the concept of science applicable in different fields and manageable in different historical periods. We expand on this concept by noting that science is, like any other set of behavioral rules, a product of the history of the culture in which it was given form and recruited its followers. For science to insert itself as acceptable to cultural ideologies, socio-cultural problems must be addressed. For example, in the late thirteenth and early fourteenth centuries in Europe, the aspiring scientist lived with the recognition that the church claimed and held sway over all intellectual territory. In Italy, Cecco d'Ascoli (Francesco Stabili of Ascoli Piceno) was one such, though in his times a scholar such as he was an astrologer/astronomer, poet, polemicist and writer. Because he was a student of nature, he was also thought to be a *mago* (magician). He and others like him wrote that all could be learned by studying nature and the stars. There could be no miracles, and no possibility of them, he affirmed. Because such naturalist-astronomers condemned much religious dogma as false magic, they are considered the forerunners of modern science. Cecco went further than most in predicting a language for the new science; in one of his few remaining works *L'Acerba* he emphasized that the new nature-science-philosophy would not be adaptable to the poetic joyous talk of the poet, but rather would require a hard and bitter language. As a contemporary of Dante, he thought that writing deep thoughts in poetic form (Dante over-poeticized, in his opinion!) would inhibit knowledge of natural laws and events (4). Cecco had studied extensively, first in Ascoli, then in Salerno, then Paris. Enduring friendships grew out of these scholarly encounters, and achieving their goals required another use for the creative imagination. He belonged to a sect considered heretical; along with Dante Alighieri, Cino da Pistoia, Boccaccio, Francesco Barberini and others, they formed a closed secret society known to themselves as "*i fedeli d'amore*" (those faithful to love), i.e. faithful in love with the goal of learning about nature and natural laws. They used the metaphor of romantic love to describe the passion of the intellect and pursuit of knowledge in the universe. To exchange and further their ideas they used code words and symbols and wrote in poetic form (5). Cecco d'Ascoli, the most outspoken member of the society, was burned at the stake along with most of his writings in Florence in 1327. These and a few others of the time provided the creative, imaginative aspects of the new science. It took time for their activities to have the impact they desired. Giordano Bruno, after having lived many years in England and France, was burned at the stake in Florence in 1600 because of his acceptance of the Copernican hypothesis of planetary motion. Without these revolutionaries of thought and their courage one wonders how others could have been led to incite a great social movement. By the mid-seventeenth century, those who studied natural subjects had learned to value measurement and the importance of learning how things and events related not only to ourselves as humans, but also how things and events related to each other.

How do scientists "do science" and how do they think they do it? Broad and Wade suggest that the myth put forth by philosophers of science that science is a completely logical process has had a lasting influence on the perceptions of scientists about their work and on how they view themselves (6). As a consequence, scientists often indulge in the "dismissal or suppression of any non-logical elements they might be aware of, denying the scientific process a major element of how it proceeds" (7). Our experiences lead us to appreciate these insights. To learn in classrooms that you will be unfailingly practicing an ideal and constantly led to believe that there is a disconnect between this knowledge and what goes on around you in the real world is another dimension of our definition of an "ideology". To disregard this problem and to ignore the historical records of how scientists have thought and worked over the centuries, is, as Kuhn has said, "deeply and probably functionally ingrained in the ideology of the scientific profession" (8).

How does much of what we might term "the total scientific product" come about? Some social scientists have studied the process in a particular setting over a short period of time; see, for example, "Laboratory Life; the Social construction of Scientific Facts" (9). We have been observing the process over many years in our work in cultural, medical, and legal settings. The French anthropologist Lévi-Strauss, in his groundbreaking "The Savage Mind" (10) gives us his view of what could be called science among so-called primitive peoples. He describes activities which we would not call scientific, i.e. totemic reasoning, plant collecting for healing and uses of plants that include mythic and symbolic elements, and certain activities to assure that hunting game will be successful. But he points out that these activities would fall under our categories of observation, classification, and cause and effect, whether we see them as "correct" or not. For in trying to understand them, it is we who do not know when something is literally meant, or when it is metaphorical yet still meaningful. "Native classifications are not only methodical and based on carefully built up theoretical knowledge. They are also at times comparable, from a formal point of view, to those still in use in zoology and botany" (11). He describes native science as comparable to what the French call "*bricolage*", a term which is not precisely translatable into English. In the art of *bricolage*, one has a problem and "tinkers" with it with whatever is at hand: in this sense it emerges as a process or an "art". Lévi-Strauss contrasts the native with our scientists who, in his words, "never carry on a dialogue with nature pure and simple, but rather with a particular relationship between nature and culture definable in terms of his particular means, period and civilization, and the means at his disposal" (12). We do not disagree with this statement, except to add that this is also true of all "primitives" contemporary and past, and to suggest from our observations that scientists are indeed "*bricoleurs*" approaching challenges with successes of the past and paradigms of their immediate past experiences, casting about for ideas wherever they are to be found, searching for new grammars for what is the same and what is different. Complexity does not alter formal similarities. *Bricolage* has taken us a long way. Intuition need not be disparaged; it can be described as stored knowledge from experience that has yet to be classified.

There are valid ways of knowing phenomena in the world other than the "scientific" way, and, for specific purposes, these ways are equally valid. We have only to consider the notion of axioms (Euclidean) in geometry and to know now that some axioms are no longer "external truths" n.b. Riemann's "there are no parallel lines" geometry, for example, as well as Bolyai's and Lobachevski's introduction of the notion that "there is an infinite number of straight lines which do not intersect...". In such geometries the Euclidean "planed" world is not all there is, since one can work with curved space. A hypothesis is always speculative - never "scientific". Even in mathematics, we have "confidence limits" about assuming that from the truth of some axioms, certain certainties follow. After Gauss and Riemann brought us to the idea that, "there was a third basic type of geometry and there were infinitely many 'mixed' geometries...it was high time that mathematicians stopped proving things and considered just what they were doing" (13). Furthermore, Gödel proved that in mathematics no formal system of axioms could be, at the same time, consistent and complete. A consistent system includes undecidable propositions, while a complete system can prove everything, for example, that ‘p’ is not ‘p’ (14). Truth, then, is "function of knowledge that is necessarily either incomplete or illogical. Such a truth is of little value and so one is forced to turn from 'truth' to 'belief', grounding knowledge in experience" (15).

Causality appears to be simple in a courtroom. A judge or jury arbitrarily decides whether to look for the essential cause, or the substantial cause, the sufficient cause etc. to assign blame and retribution; it appears to be a cultural/moral judgment. It appears to be non-arbitrary when science is involved - but is it?

Do we always know the cause of shifting states, recurrence of a previous state, or are we almost always somewhere between statistical laws and a process of reducing the variables? (16). A classic entrenched view of causality in science is that it is invariant, necessary and sufficient, and contiguous, but this view has been found debatable, (especially, for example, when analyzed in terms of modern physical theories (17). "Believing", it would seem, has been stored in the hard drive of the continuing human intellectual endeavor.

In summary, there are uncertainties in science because we deal with necessarily incomplete data in a probabilistic manner. On the other hand, uncertainty in legal systems, in any culture, has to do with human beliefs and behavior, often expressed through agents of change who construct, deconstruct or reconstruct aspects of legal reasoning. And also, we must introduce "chance". As in any "game", a person or a group has to guess the strategies of the opponent and then calibrate their own responses in a probabilistic manner, knowing that the outcome is uncertain.

NOTES

1. Ludwig Fleck. *Genesis and Development of a Scientific Fact.* Chicago: University of Chicago Press, 1979, p.8.
2. The term "soft science" includes the "social and behavioral sciences which claim an objective understanding of human nature". See David L. Faigman: "To have and have not: assessing the values of social science to the law as science and policy". *Emory Law Journal.* 38:1005-1008. 1989.
3. Philip Frank. *Philosophy of Science: the Link between Science and Philosophy.* Englewood Cliffs, N.J: Prentice Hall, 1951.
4. Lola Romanucci-Ross. *One Hundred Towers: An Italian Odyssey of Cultural Survival.* New York: Bergin and Garvey 1991, pp112-115.
5. Luigi Valli. *Il Linguagio Segreto di Dante e dei Fideli d'Amore.* Rome: Optima 1928.
6. William Broad and Nicholas Wade, op. cit. (Chapter 1).
7. Broad and Wade, op. cit. (Chapter 1) pp126-127.
8. Thomas Kuhn. *The Structure of Scientific Revolutions.* Chicago: University of Chicago Press, 1970. Second Edition.
9. Bruno Latour and Steve Woolgar. *Laboratory Life: The Social Construction of Scientific Facts.* Sage Library of Social Research. Vol. 80, London, Sage Publications 1979.
10. Claude Lévi-Strauss. *The Savage Mind (La Pensée Sauvage).* London: Weidenfield and Nicholson, 1962 pp. 62-66.
11. Lévi-Strauss, op. cit. p. 43.
12. Lévi-Strauss, op. cit. p. 17.
13. John G. Kemeny. *A Philosopher Looks at Science.* New Jersey: D. Van Norstrand, 1959 p. 124-125.
14. Raymond Wilder. *Introduction to the Foundation of Mathematics.* New York: Wiley 1952, pp. 256-261.
15. Lola Romanucci-Ross and Daniel Moerman. "The Extraneous Factor in Western Medicine" in *The Anthropology of Medicine: From Culture to Method.* L. Romanucci-Ross, D. Moerman and L. Tancredi, Eds., Westport, Conn: Bergin and Garvey, 1997, p. 352.
16. Frank, op. cit.
17. Ernest Nagel. *The Structure of Science: Problems in the Logic of Scientific Explanation.* New York: Harcourt and Brace, 1961.

CHAPTER 3

FROM MYTH TO LAW

Every society has a mythology surrounding the origins of rules of conduct and systems of law, revealing that law is not based on experimental data. Such myths most often begin with a postulated relationship to a sacred presence or personage in the universe that might be viewed as "contractual". Cultural groups may or may not have sacred places and objects, but in all of them the foundations of rules of conduct and proper relationships to persons, animals and things cultural or natural, reside in the contract with the sacred power. The contract includes agreement about good and evil (crime and punishment), group membership (allowed and disallowed, with exclusionary provisions), marriage rules, sexual taboos, sexual access and denial, punitive measures meted out through illness, death, or the many forms of loss, as well as some advice for conflict resolution (1).

All this would seem to argue for a universal recognition that tenets of law are not evidence-based but rather appear to be a common sense approach; thus, truths about inter-personal or inter-class relations are never "self-evident", even though we may choose to "hold" that they are. From any such evidentiary assertion, corollaries flow - always justified by the first assertion and basic belief. These having been established, we then find justification in "legal precedent", creating a self-contained structure of ideologically driven articles of faith translated into social action. These will be "held" until there is cultural pressure to revise what we hold to be self-evident (2). This does not mean, however, that such a belief system is less "real" than empirical truths. For beliefs determine behavior and have the impelling force of a "reality" constituted of cascades of proof and validity, and because outcomes are usually predictable, beliefs have a good chance of being reliable.

A legal system can be thought of as a mathematical system, in the sense that we speak in terms of postulates and premises. In practice, a legal system is non-mathematical, for the most part, often ignoring strict adherence to precedent, postulates, and principles. Concepts of efficacy, efficiency, implications for administration of justice and, most fundamentally, social beliefs, political exigency, and chance (choices of judges and juries) come into play, overwhelming the structural framing of problems and their solutions.

What are some of the values, which are expressions of an Ideology, that underlie the American legal system? First, we believe that justice is attainable, can be defined, may have to be distributive in nature, and that all are equal under the administration of the law; also, justice may be appealed, and appeal allows for corrections--reducing the role of "chance". We have our specific notions of causality: that there are direct and indirect causes; that "motivation" (a sort of cause) can be imputed by another, and that this too can be contested. We have notions of redress: that money can be equated with life, or happiness, or health; also, one has the right to seek redress and the state has the right to seek redress for an individual or group in concepts of property or space (personal and other). We have the notion of contract and that a society can enforce it. We stipulate that one must be judged by a jury of one's peers, and that one has the right to a public trial; that one has a right to legal representation that must be provided if one cannot afford it. We mandate that our court systems handle various appeal categories, e.g., Federal Courts primarily deal with constitutional matters and issues of jurisdiction and large monetary awards (3).

How does this conceptual model of our legal system drive us to address our ideologically driven social action model? Even mechanical models, in which the a priori nature of a system of thought is often explicit, depend very heavily on contexts and the introduction of new assumptions for problem solving, as Gödel noted. In Gödel's Theorem, any set of mathematical definitions and postulates gives rise to further questions that cannot be answered on the basis of those definitions and postulates, i.e., new questions can be answered only by assuming another postulate which then gives rise to new questions requiring yet another. He further noted that there is no lower context which is not provisional and such a context attains meaning only in the sense or measure that it gives meaning to an upper context. In legal systems, reasoning resembles a mathematical system in that there is a built-in logic: things have to be resolved the way they are because they are set up to reach certain preordained "correct" solutions to problems as they are stated. Change occurs only when questions arise which cannot even be addressed, let alone solved, without new postulates. These will depend on new agreements about principles concerning rights and duties and relations that are to obtain between people and property, people and other people, etc. The stylizer for new postulates are to be found in culture change (social, political, economic, religious, or combinations thereof).

Cultural factors also enter the legal system through the insight of the observer in establishing "facts" and values. Here again, we can include the mathematical notion of "complete" and "incomplete" systems, for when a judge makes a decision it creates law which may be going outside the system; when an appeals court overturns it the system is "re-stabilized", keeping the system closed (i.e. "complete"), for a time at least. Let us bear in mind that we are theorizing about "moving scenes"; we earlier noted that mathematics is also a "moving scene", exemplifying this through the evolving new geometries. Both the evolution of "rules of evidence" and Gödel's view of changing contexts and postulates fit this concept.

In practice, law is on the brink of being a social science, since the objective is primarily conflict resolution rather than a continuous exercise in knowledge acquisition. Therefore, law must rely not only on factual data and interpretation, but on overarching principles of effectiveness, impact on the social fabric, administrative simplicity and efficiency in the setting of operative decision-making. Thus we see an evolution from Frye in 1923 to Daubert in 1993 (4) for rules about which evidence is admissible; this was an evolution predicated in large part on cultural pressures during changing historical periods for achieving specific policy objectives. "Causation" may vary from case to case. In some, a "but/for" causation (necessity) is essential as the basis for ascribing responsibility whereas, in others, a "sufficient" causation (without the requirement of necessity) would suffice. In still other cases, the causal factor may be a "substantial" element (though not necessary and/or sufficient). If for economic, "efficiency" or social reasons it is desirable to find a cause even when the causative factors are only weakly linked to outcome, then a less demanding test may be acceptable.

In mathematics, a proposition can be deemed true or not true because the manner in which it is true can be demonstrated and conclusions can be causally understood through general agreement on the axioms. In the legal arena, from the appeals court through the Supreme Court, the law in action concerns itself with concordance of judges on legal axioms. Concordance or lack of it makes it possible for the law to adapt to changing cultural values.

Law and medicine, both disciplined thought systems with axioms, have different ways of reasoning about the world, are both concerned with rights of individuals in society--the right to justice and the right to well being. Both have conscious thought models and an awareness that these can be modified. In simpler societies medicine and the law are one, in a literal sense. For even in these societies, although medicine is the art and science of cure, prevention of suffering, and preservation of health, it also includes magical power over nature and over other people through rituals and charms, and the "magical" laws that govern them. "The law", concerned with rules of conduct enforced by a controlling authority with sanctions and injunctions, is considered binding by members of such societies; members who respect the rules are rewarded by good health and when they do not respect them, they are punished by illness (reversible or irreversible) (5). And herein lies the goal of their medico-legal machinery: health care and social control are the lynchpins of social survival. For the Native American Indians, good medicine meant all of those things that maintained social harmony and individual health. The body and the Body Politic were subject to the same laws for optimal personal and social functioning. In the Admiralty Islands, as we noted, illness, curing and healing were always socio-moral events. In illness one found the accusation or the indictment, in diagnosis the sentence and perhaps the punishment, and in the cure the proof that the guilty person had set all aright.

In passing, one might also note, that biological and botanical knowledge are among the most important types of information needed for survival in any culture. Herbal medicines are not patented in most cultures. By the 19th century, we in the Western world had invented patents, copyrights, "trade secrets" and trademarks. We have now decided that Ideas are property. As in earlier historical periods various disciplines separated and claimed either body or mind, in our times designated medical and legal professions claim biological or legal knowledge as property.

NOTES

1. In "primitive" societies such relationships, ensuing "contracts" and emergent legal systems are expressed in terms of a totemic relationship: it is thought that members of a lineage descend from a totem (animal or plant). Western minds have long struggled to understand and interpret "totemism", a label they themselves (the scholars) have put on this cultural phenomenon. Anthropologists accept that it is characteristic of many primitive cultures to classify lineages within their groups as descendants of a totemic founder, an animal or plant. Some social scientists viewed it as an expression of social order (Durkheim); anthropologist Lévi-Strauss defined it as a method of classifying human groups by using natural world models as labels, and perhaps also denoting characteristics that others might have attributed to this group (Lévi-Strauss 1962). Anthropologist Theodore Schwartz views the concept of totemism as involving many processes both social and cultural, including "identity-seeking" with political overtones. Within an area of differing cultural enclaves, a totemic lineage interacts with others in many interlacing ways, including proprietorship of cultural differences. (Schwartz, T. 1995).
2. Exemplary is the American Declaration of Independence of 1776. After September 11, 2001, President Bush declared an all-inclusive war against terrorism; for this decision there was virtually unanimous support in the country. Some of the administrative provisions that followed for implementing this, however, troubled some Americans and were viewed as threatening constitutionally guaranteed rights. Such historical moments of crisis which pit worthy goals against worthy goals are a dramatic instance of possible sea changes in what a people hold to be obvious "self-evident" truths.
3. For an interesting typology of "rights", see Haack pp. 14-17.
4. Frye v. U.S. (Frye 1923) made it requisite that an expert's testimony to be introduced in court be founded on methods and theories that were generally accepted by the scientific community. In 1975 admissibility was reconsidered and codes were changed: expert testimony would now be allowed in the court if any scientific, technical or other specialized knowledge would help the trier of fact to understand the evidence. Mainstream acceptance was not the preeminent factor (Federal rules of evidence 1975; see also Giannelli and Imwinkelrod). In Daubert v. Merrill, the court, in its decision also added a commentary on the problem of expert testimony and scientific evidence. Judges were, henceforth, to be the gatekeepers for what was minimally acceptable as such as well as notions of standards of relevance, reliability and validity. It did allow, however, that judges might need a short course to understand the scientific method. In 1997 the court allowed judges to bar "junk science" and gave judges the power to opine about and bar what they considered controversial scientific evidence. In addition, appeals courts were not to go about second-guessing trial judges who excluded scientific evidence that came from "shaky studies" and dubious expert witnessed. (General Electric Co. v. Joiner, 522 U.S. 136, 144 (1997). However, in 1999, the court decided that technical experts could be admitted as expert witnesses since some forms of such expertise are indeed grounded in scientific principles. (Kumho Tire Co. Ltd. et al v. Patrick Carmichael, et al., 526 U. S. 137; 119 S. Ct. 1167 (1999).
5. Lola R. Schwartz (aka Lola Romanucci-Ross). "The Hierarchy of Resort in Curative Practices: The Admiralty Islands, Melanesia". *Journal of Health and Social Behavior.* 1969 10:201-209.

CHAPTER 4

THE ROMANCE OF SCIENCE AND MEDICINE

The word "science" is polysemic, which is to say that it has a spectrum of meanings to speaker or listener and writer or reader, whether scientist or layperson. Thus, the semantic domain of "science" encompasses many concepts that are selectively shared by many.

The scientist is believed to have attributes such as dedication to truth seeking, to revealing reality through observation, experimentation, and analysis. Methods and results deemed publishable by one's peers must have the qualities of reliability and replicability. The accoutrements of science are imbued with symbolic value: the white coat (in the laboratory or the medical clinic), the various tools and equipment for measurement, and the legendary, long training periods (pushing the acolyte to the limits), which presumably exorcise all other possible emotional attachments that might impede perfect outcomes.

Because of these attributes, science is considered a calling somehow different from other disciplines. It should be, it is thought by some, less vulnerable to lay criticism, less accountable, beyond attack. It is seen (particularly by its practitioners) as a self-correcting discipline, on auto-pilot toward total revelation of all we could ever want to know and understand.

For many scientific thinkers, there is an architectonic of ways of knowing, with natural scientific inquiry at the apex supported by heavy or exclusive reliance on data (often ignoring the metaphysics of how those "data" are gathered and arranged). Other scientists, because of predilections of personal philosophical reasoning, favor deductive or inductive approaches to problems of discovery. Few would recognize or ever admit that ideology or beliefs have anything to do with the nature of a "fact". Furthermore, "Science" is considered above and outside culture (culture-free).

How did we arrive at the notion that science is value-free, untouched by culture, unsullied by concerns of personal goals and the need for social approval? Broad and Wade noted that "the analyses of fraud sheds considerable light on how science works in actual practice" (1). Peer recognition of personal achievement and accolades from society are motivators and stylizers of scientific research, even though there are many scientists who love their work and find it be sufficient reward (although we have yet to meet more than one or two who don't hope for more).

Throughout European history, scientists have had to insist on freedom from social, political and religious pressures in developing their field. Some, such as Giordano Bruno, have proved their point by allowing themselves to be burned at the stake. Others, such as Galileo, muttered a rejoinder (as an aside, after a recantation on heliocentricity) "*eppur si muove*" (nevertheless it moves!). The intellectual descendants of these science-culture heroes have come to believe that the situation is actually what they wish it to be: culture-free, values-free and untainted by ambition.

But we find ample evidence that science needs an aesthetic, i.e., imaginative poetic creativity for its advancement. We know that there are power hierarchies and reward systems in all scientific enterprises. It is recognized within areas of scientific endeavor that there is no unanimity on choice of method, preferred explanation, or most promising avenue for further research. And it is well established that increasing the number of one's publications enhances opportunities for promotion and for obtaining further research funding.

An image shared by many (but usually not by practicing scientists) is that of the great man, who through a *coup de foudre* makes The Discovery, and to whom (in this Kuhnian moment) all is revealed! Although this is not the way discoveries usually come about, for a plethora of reasons it is often represented that way. Of many examples, from the "insights" of Darwin on the theory of evolution (2) to relativity theory and its many precursive thinkers, we may consider the "structure of DNA" discovery. This Nobel prize-winning "insight" has been meticulously documented by Judson in "The Eighth Day of Creation" (3) and in a remarkably candid personal account by James Watson (4). We recommend this example, since both books are a good read; they deal with a subject that is close at hand, with most participants alive and available for interview at the time of writing, and all of the work still easily accessible. The insight that created the model of the structure of DNA was itself partially but significantly structured by data obtained with (and sometimes without) the knowledge and permission of the scientist who created specific data (such as those obtained by X-ray crystallography [Watson]), as well as from the work of many scientists whose prior and ongoing experiments provided crucial pieces of information.

FROM THE LIFE AND TIMES OF RENÉ DESCARTES

The successes of "method" in the sciences were dependent upon the division of "physical" and "mental". Prior to that, as we indicated earlier, during the times of Giordano Bruno and Cecco d'Ascoli, among others, the church was an impediment to the pursuit of a scientific method. The mind-body problem concerned Descartes who tried to understand how something non-spatial (such as "thinking" or thoughts) could be related to spatial matter. It was in his writings and his recognition of this problem that, in the 17th century, we find the roots of modern science and medicine.

Descartes' influence on the empiricist philosophers is especially notable; for they insisted that only making explicit exactly *how* we know something should be the starting point of any philosophical endeavor. Logical positivists later asserted that the meaning of a statement can be found in its verification (Schlick). In other words, a concept is synonymous with the set of operations that determines its applications (Bridgman). Post-empiricists such as Husserl tried to recapture the Cartesian spirit; in the words of Williams (1967:354), "the problems posed by Descartes' dualism remain at the heart of much contemporary philosophical inquiry (the work of Gilbert Ryle and Ludwig Wittgenstein, for example) being aimed directly at what are still very powerful Cartesian conceptions".

In recent times, the philosophical school popularly known as "Kuhnian" has granted a more significant role to thought itself in the process of knowing.

According to Kuhn (1970), one searches for the new paradigm that will revolutionize the manner in which a problem is phrased as well as "solved". Once introduced and accepted, however, the paradigm tends to stabilize thought, thereby reinforcing the new belief system, so that it is likely to remain unexamined.

To contrast the views germane to our discussion, in the 17th century Descartes could affirm the possibility of knowing the "truth" without a single doubt, based on his own *experience*. He "knew", for example, that blood was pumped through the body because the heart was hot, which caused the blood to expand and to course out through the arteries Descartes warned his readers against casual criticism of his work:

> "I would like to warn (critics) that this movement (of the blood) that I have just explained follows as necessarily from the mere disposition of the organs which may be observed in the heart by the eye alone, and from the heat that one can feel there from one's fingers, and from the nature of the blood which can be known by experience, as does that of a clock from the power, the position and the shape of its counterweights and its wheels" (Descartes).

Similar properties caused similar behaviors regardless of context, an inference constituting a sort of naïve realism. For Kuhn the paradigm (made up of categories, relationships, and decisive examples) structures the view of truth that the scientist discovers. When the paradigm changes it is "rather as if the professional community had been suddenly transported to another planet where familiar objects are seen in a different light and are joined by unfamiliar ones as well...nothing changes but the view" (Kuhn 1970:111). Thus, the Kuhnian view emphasizes the changing nature of knowledge as viewed through new paradigms, whereas the Cartesian view builds persistently on past observations and perceptions.

TO OUR OWN

In an earlier section as we considered "matrices of meaning", it was asserted that in both the professions of law and medicine the training period (i.e. the enculturation process) provides the professional with a new language, new ways of relating to others and new rituals. Most importantly, however, the aspirant professional has to learn new ways of thinking (although most medical students or interns would not characterize it that way). While lawyers thrive on ambiguity, finding it productive and creative, the medical professional has low tolerance for it. One could arguably conclude that for either profession the preferred stance towards ambiguity has survival value; for the physician, however, it may have survival value for his patient as well. Our purpose is not to pass judgment on values but rather to consider how such pressures influence modalities of reasoning—how a professional "thinks". Physicians seek order through classification; both order and classification are constructs imposed on the "real world", even though we tend to think that the order is real and the classification describes it. The physician's order (read classification) is the basis for his general procedure: to measure, to compare, to predict. Values of mastery, order and power are axiomatic; they need not be thought about. They precede thought; they provide the scaffolding for it. They are good, in and of themselves, moral virtues. When these issues *are* considered, they are transformed into virtues of the highest sort as they were for the ancient Romans who equated knowledge and virtue.

ON VALUES AND SOME RESEARCH MODELS IN WESTERN MEDICINE (10)

Contemporary ideology in biomedicine includes the following values: first, causation is temporal, reductionist and essentially metonymic, that is, the part represents the whole and/or cause is taken for effect and effect is taken for cause. The metonymic style looms in importance as it forms the primary structure for treatment. One "treats" the "cause" of illness, not the "symptoms" of illness. This treatment is usually conceived to be allopathic, which (in contrast to homeopathy) is generally the consequence of another major belief in biomedicine: disease is natural while, in general, healing is culturally determined. A general corollary to this belief is that disease must be treated, even if effective treatment is not available. Culture must at least contest with nature, even if the game is already lost. Moreover, patients who do not respond to standard treatments (morphine for pain) were not sick in the first place; they may be classified as "hypochondriacs". Similarly, people who respond to inert treatments (placebos) were not really sick, either. A study carried out in a Texas hospital showed that house officers and nurses considered some patients to be "problem cases" because they did not respond to standard treatments, leading to the hypothesis that they were not really sick. When they responded favorably to placebos, this response was considered verification of the hypothesis (Goodwin et al).

Similarly, there is a standard reaction to occasional cases of "spontaneous remission", in patients who have been refractory to treatment; often, this is explained by saying that the condition was probably misdiagnosed in the first place. These sorts of phenomena—ineffective drugs, effective placebos, spontaneous remissions—are generally considered "impossible" given the ideological principles under discussion.

Of course, the placebo, or biomedically inert substance, also helps eliminate bias in research protocols, although a beneficial placebo effect does not fit the Western master paradigm:

> Many papers have demonstrated the importance and magnitude of the placebo effect in every therapeutic area. Placebos can be more powerful than, and reverse the action of, potent active drugs. The incidence of placebo reactions approaches 100 percent in some studies. Placebos can have profound effects on organic illnesses including incurable malignancies. Placebos can often mimic the effects of active drugs. Uncontrolled studies of drug efficacy are reported effective four to five times more frequently than controlled studies. Placebo effects are so omnipresent that if they are not reported in controlled studies it is commonly accepted that the studies are unreliable. [Shapiro 1968:58]

Brody argues that the physician of our time cannot deny the placebo data, "but he can adopt an attitude towards it of exclusion, that is, labeling the placebo effect so that it can be readily recognized and thus excluded from research" (Brody 1980:27).

We mentioned Descartes above, and it is appropriate to refer again to the man and his period (the 17th century) since in the times and in the man we recognize the foundations of modern scientific thought. Descartes respected his teachers but stated that only mathematics had ever given him "certain knowledge" (Edwards 1967:344). He stressed quantitative measurement and experimentation as well as taxonomy and calculable order. Descartes accepted the notion of circulatory movement of the blood but not the independent contraction of the heart as the driving force of the system. More inclined to mathematics and deductive thinking as discussed earlier, he looked for another mechanical cause of blood displacement: its "heating" and subsequent expansion as it entered the heart (Snellen 1984:22).

William Harvey took up the mistaken idea of "heat" but also *calculated* the total volume of blood in the circulation and estimated the output of the pumping chamber of the heart per beat and per minute. He therefore was able to conclude that the blood must recirculate, driven by the heart (Harvey 1628, trans 1978). This ideology, with its commitment to observation, measurement and experiments, led not only to some false starts (through Descartes), but also to productive work (through Harvey) that led to later "knowledge explosions". However, important for our purposes, unexplained ideology still thrives in the research paradigms of our day. We have typified the much valued canons of correct "scientific" behavior and give several examples:

A *randomized clinical trial* is the study of the effects of an intervention in a population expected to experience abnormal events with a given frequency. The trial is undertaken in a population sufficiently large so that the treated and control groups have the same characteristics; neither the investigators nor the patients know whether they receive placebo or active medication or some other intervention such as surgery (i.e., they are "blinded"). The goal is to determine whether or not a specific intervention over a period of time will reduce the abnormal event rate in a statistically significant manner. As an example we refer to the "Coronary Artery Surgery Study" known as CASS (CASS 1983). In this large study, patients were randomly assigned to either customary medical treatment for their mild to moderate chest pain (angina pectoris), or to coronary artery bypass surgery. After a (mean) six-year follow-up, although some of the end-points were of interest, no significant difference in survival was found between medical and surgical treatment (for surgeons, an unexpected outcome). Later studies showed survival benefit in several patient subgroups, but our purpose here is to illustrate that clinical trials can have pitfalls. Thus, while the randomized clinical trial is the ideal method for assessing therapeutic interventions, even when everything from the complete, carefully collected baseline data and all statistical pitfalls are taken into account, there is still a likelihood of "chance" differences achieving statistical significance when multiple subgroups are analyzed. In a study carried out at Duke University, 1,073 coronary artery disease patients were randomly sorted into two groups; there were no differences in treatment between the two groups of patients. There was no overall difference in survival between the two groups, as one would expect. However, in a subgroup of 397 patients with three-vessel disease and abnormal left ventricular contraction, survival was significantly different in the two randomly sorted groups. Multivariable adjustment procedures attributed the difference to the combined effect of small differences in the distribution of several prognostic factors. On a univariate basis, the "treatment" (that is, randomization) might appear to be a significant factor in the survival of this subgroup (x^2 = 5.4; p .025). When the variables were considered jointly, the "treatment" effect became nonsignificant ($x^2 = 2.4$; $p = NS$). In addition, in another subgroup a significant survival difference was not explainable even by multivariate methods. In this case of patients with three-vessel disease, abnormal left ventricular contraction pattern, and no history of congestive heart failure, the respective three-year survival rates of the two randomly differentiated groups were 60 percent and 80 percent ($x^2 = 10.0$; p .01). The authors of this study caution that clinicians must exercise careful judgment in attributing results to efficacious therapy, as they may be due to chance or to inadequate baseline comparability of groups (Lee et al. 1980:508-515).

Indeed, physicians inevitably use "judgment" when they prescribe treatments; this is to say that "science" (the clinical trial) is tempered by "knowledge grounded in experience"; these are then combined into what physicians call "clinical judgment" (which is further addressed below). The randomized clinical trial has had an important effect in medicine. But the design of trials necessarily reflects the assumptions and expectations of the investigators and may provide convincing demonstrations of "obvious" but incorrect notions (cf., Descartes 1968[1637]:68). Statistical significance is no assurance against design error.

In models of intervention and observation one finds what there is too much of or too little of, and supplies it or changes the balance. In many instances this has caused great relief to the patient, but often at the expense of long-term effects that are noxious or lethal. Many asthma sufferers have ended up with hypertension (generally treatable) after eight to ten years of treatment with steroids. Those taking the phenothiazines (such as thorazine) have acquired irreversible tardive dyskenesias. In such cases, the effects of drugs on biological systems through time are not taken into account. There are other examples: use of diuretics can lead to elevated uric acid levels and development of gout. Radioactive phosphorous used to treat polycythemia vera can lead to leukemia. Cyclosporin used to prevent rejection of a transplanted organ may cause cancer.

Adverse interactions of drugs given at the same time or in close sequence (often by different physicians unaware of or unconcerned with one another's treatments) illustrate possible outcomes of ignoring the time factor, the patient's history, and the synergistic effect of drugs as they affect or influence one another. Effects can also occur by combining drugs with certain foods.

Although Withering described the therapeutic values of foxglove extracts 200 years ago, it is only recently that some studies have provided a demonstration of mechanisms of important extracardiac effects of digitalis glycosides, such as slowing of heart rate, rise in blood pressure, and other actions mediated by the central and peripheral nervous systems (Longhurst and Ross). Extracardiac "side effects" had long been disturbing to patients who experienced nausea and vomiting, abdominal pain, gynecomastia in men and breast enlargement in women, vision disturbances, headaches, seizures and coma. Studies such as these, revealing the mechanisms of "drug action", should lead physicians to a revisionist view of the simplistic notion that such actions can be considered in terms of "primary" and "secondary" effects, as is currently the case.

We consider here more briefly a few other aspects of contemporary medical research and practice where extrinsic factors play an important role limiting and constricting understanding: the use of animals and analogic thinking, models for sane/insane behavior, and medical diagnosis.

Investigations that use animals provide us with an interesting example of the uses of metaphor in medicine. Further, they indicate the difficulties that can arise when metaphors are taken literally. Dogs, pigs, and baboons (among others) are widely used in cardiovascular research. At times, the pig provides a better representation of humans for, unlike the dog, it has a large right coronary artery, and few coronary collateral vessels. Sometimes, however, the dog is a good model because it has more collaterals, and a number of people with coronary heart disease develop significant collaterals. The baboon is a better hemodynamic model of humans because of its more upright posture. One must, then, match the animal to the problem at hand. These considerations have been noted in the scientific literature (Crozatier et al; Sanders et al.; Tomoike et al.), but they have generally been ignored as investigators go through the ritual of scientific presentation, confusing the metaphorical model for the object of their ultimate concern. This is often overshadowed by the "naïve realism" (if we may so call it) in some public attitudes that there are *no* similarities between any animal species and the human species.

Attachment to animal models of disease transmission has sometimes impeded careful inquiry and more correct epidemiological explanations of the spread of certain diseases, and a study of historical records with modern research techniques and current medical knowledge can yield surprising results. For example, transmission of "the plague" was attributed to the abandonment of the infected and dying *Rattus rattus* by its resident flea (*Xenopsylla cheopis)*, which then regurgitated the plague bacillus (*Yersinia pestis)* into its human victim. But a consideration of such factors as seasonal occurrences of the disease, the number of cases per household, comparisons with the presence of the black house rat in other parts of the world, and the sudden disappearance of the plague in Europe has led to a broader hypothesis. Ell's research (Ell) has emphasized interhuman transmission as well as several varieties of "the plague". (Medieval physicians knew the plague quite well and described variations in great detail.) Those who wrote treatises concerning explorations of the occurrence of the plague had rested comfortably on the rat-flea-human mode of transmission, which, though correct part of the time, stifled further investigation for a long period.

The sane/insane model of human behavior typifies the procedures by which mental states and behaviors have been medicalized, traditionally rendering patients helpless as their caretakers decide on medical and surgical interventions. But modes of communication were not part of the general ideology of the biomedical practitioners and were considered extraneous factors. It was only after the legal battles of the civil rights era—the 1950s to the 1970s—that mentally ill persons were given a "say" over their patient status (Tancredi 1983). Indeed, many health professionals now recognize these factors, but still consider them extraneous, intrusive and destructive (Radine).

The confluence of the psychiatric model with some features of the general diagnostic process can provide us with an interesting field for the compounding of errors. In the Soviet Union during the 1960s, a system was developed for psychiatric diagnosis. Many of the most talented psychiatrists in that country were disciples of a central theory of schizophrenia, developing a definition so exceedingly broad as to include much of whatever else might be included in psychopathology. According to this definition, schizophrenia had three possible course forms: continuous, periodic and shift-like. Patients of the continuous type experienced early onset and did not improve; the "periodics" had intervals of remission during illness; and the shift-like patient was a combination of the continuous and periodic types. This was further complicated by the additional diagnostic criterion of "severe" or "mild" for any type.

What was (or should have been) troublesome was that many "patients" of the "mild" type would not have been seen by any psychiatrist at all in another milieu (Hite; Rollins). Also, the International Pilot Study on Schizophrenia of the late 1960s and early 1970s noted that two of the nine centers reviewed for diagnostic activities did poorly: Washington and Moscow (World Health Organization 1973). The goal of the computer program in this study was to rediagnose patients originally diagnosed as schizophrenic at various centers, using the center's own data on the patient. Moscow's Institute of Psychiatry and Washington did very poorly for different reasons, however; a large percentage of Moscow's diagnosed schizophrenics were reassigned by the computer to neurotic and depressive categories rather than the psychoses, whereas the Washington Center had many of its patients reassigned to the psychotic categories. Obviously one center "over-diagnosed" and the other "under-diagnosed".

Some have noted that psychiatric diagnoses have provided a solution to certain human problems (in some cases a "political" solution) (Basaglia 1980, 1981). Reich has noted that such diagnoses are also used to *reassure* all of us; he gives the example of a researcher, Dr. Summerlin, who inked the skins of his mice to make it appear that grafted skin had "taken", while others could not repeat his experiment with his dramatic results. It was important that the public not be led to think that research in general was a fraudulent activity, for great sums are donated to the research enterprise. The investigating committee "found" that Summerlin's behavior involved self-deception and aberration. Even Dr. Lewis Thomas, president of the Sloan-Kettering Cancer Center informed reporters that "the fraud in this work was the result of mental illness" (Reich 1981). The same explanation was put forth by those surrounding Dr. John Darsee's fraudulent research in a Harvard Medical School Laboratory (Knox 1983). Such diagnoses appear to be self-validating in the same manner that it is reasoned (with the benefit of elaborate diagnostic procedures) that a criminal "must be" insane to have committed the act for which he is undergoing psychiatric appraisal (Romanucci-Ross and Tancredi 1986).

Similarly, in medical diagnosis a variety of extraneous factors shape, form, and determine the procedure, making it far less scientific than it might be. Studies have consistently shown that medical problem-solving is a hypothetico-deductive activity in which early problem formulations partly guide subsequent data collection (Elstein).

In a review of 50 clinico-pathological conferences, the process of achieving a diagnosis was shown to have determinable steps. First, there is aggregation of groups of findings into patterns, followed by the selection of a "pivot" or key finding (in our words, the problem is metonymized). From this, a "cause" list is generated and pruned to become a set of differential diagnoses—a listing of diagnostic possibilities—from which one is selected and then validated (Eddy and Clanton). The problem, of course, is in the amount of information to be considered; one must interpret signs and symptoms to diagnose disease, but one's medical training was learned the other way around—by disease. This medical decision-making chain has an effect on the physician's "intuitive" uses of the principle of discriminant analysis, that is, how much weight is given to clinical versus statistical factors, or how knowledge is grounded in experience. The diagnostic process is further constrained because the "facts" are, of necessity, selected and evaluated in a temporal ordering. Decisions have to be made within a time frame and the presenting symptom does not always indicate that the patient suffers from several afflictions. As the physician becomes a member of the medical subculture, he or she has learned new ways of thinking, a new language, new "laws" of causality, etc, but something may have been lost, as well. For medical diagnosis is first and foremost a socio-cultural event, a product of communication between a concerned patient and a scientifically prepared professional. The diagnosis will be affected by the skills of the physician to elicit the needed information. It may seem obvious, but it is also important that the doctor listen. The physician's clinical judgment is the product of a complex function of the scanning self and his or her cognition. Through this process the many subdisciplines that converge into medical science can be combined. The physician scanning all he/she knows (and feels) is simultaneously deductive, inductive, and quantitative, while morally committed to the best outcome for the patient.

The best judgment for the patient and the professional's commitment to it has been heavily influenced by values retained through the centuries from Cartesian times to our own and have become fundamental components of our medical legacy. Seventeenth century science valued "measurement", the relating of things not to ourselves but to each other. The end of the 18th century saw the birth of Romanticism, the "transcendental", and a serious consideration of subject-object relations. But in the 19th century, what was central to medicine was clinical investigation and experimentation, critical working hypotheses, and structure, both organic and physiological. These were epitomized in the works of Claude Bernard and Rudolph Virchow. And, as Foucault pointed out in *The Birth of the Clinic* (Foucault), we had the establishment of faculties of medicine and the transmission of medical knowledge into social privilege. During this period, there was a strong beginning in the exercises of looking at probabilities in diagnoses and, presumably, becoming aware of "extraneous factors". Although Morgagni still specified diseases by points of origin, he began to find cells, nervous tissues in different organs, fibrous lesions, and lesions of serous and mucous membranes (Morgagni 1761). This began the search for "principles" about tissues. It appears to have been a period of exciting theoretical modeling and synthesis in clinical experience, as well as in research. And yet, Italian anatomists had already made essentially the same transition, from form to function, in the 16th century (Castiglioni). This historical fact demonstrates that innovation does not always overcome ideology and affect mainstream science as soon as it should and in the manner it might. Principles are rediscovered centuries later because they then fit the new paradigms. Morgagni founded the discipline of pathology, and in his classic work on the causes of disease, *De Sedibus et Causis Morborum per Anatomen Indagatis* (1761), he employed the best canons of inductive research. But he was in no hurry to publish his work, fearing criticism and rejection.

Most of the values of the Classical period of the 17th century have remained with medical science to the present day: to make one's contribution to the total body of knowledge without inspecting the interstices and intersects of the content of that body of knowledge, to dig deeper into "nature's secrets" with little regard for the consequences, to look for causality in relatively isolated systems. In biophysics and bioengineering we find elaborate metaphors from hydraulics or plumbing (when speaking of the circulatory system) or electrical circuitry (when discussing the nervous system). One can speak of neural "charges" or "discharges", and with all metaphors, these can provide insight and facilitate communication. But metaphors provide only the form for understanding; they do not imply identical formal structures or identical relations between points.

Medical science frequently structures research around linked propositions in which two or more concepts are reciprocally defined. We investigate the validity of linkages and, of course, usually find them. For example, we are often told that early diagnosis improves survival. Of course, if the mean patient survival time after contracting an illness is ten years, and if the mean time at diagnosis after disease onset is five years, the "five-year survival" statistics will be dismal. If mean time at diagnosis can be reduced to three years, the five-year survival statistics will be dramatically improved. This change, of course, has no relationship either to the effectiveness of treatment, or to the course of the disease. Aggregates of linked concepts can be made as complex as you like, but, as Gregory Bateson has indicated, the links are provided not by the data but by you.

Kant, long before Kuhn, had noted that experience itself is a species of knowledge that involves understanding (Kant). Since medical science does and must include behavior, one must apply Devereux's dictum that "the scientific study of man is impeded by an anxiety-arousing overlap between subject and observer" (Devereux 1967:xvi); thus, perception and interpretation of data are distorted, producing "countertransference masquerading as methodology", causing even more distortion. Devereux meant that such structural "anxiety" exacerbates individual vulnerability, revives more idiosyncratic anxieties, threatens to undermine major defenses and exacerbates current problems (Devereux 1967:45).

An example might be provided by a patient with an illness that causes the physician to react to his patient as though the patient were an early significant figure in the physician's life. Even more pertinent to our present discussion, however, are recently documented professional practices of certain medical scientists who publish inaccurate statements either knowingly or unconsciously and unknowingly, who describe procedures that were not followed or were followed in other experiments at other times, who "lose" primary data and/or have little direct involvement with the research leading to articles of which they are "co-authors" (see Broad and Wade; Knox). "Anxiety", apparently, can be detrimental to medical as well as social scientists, nor is social science the only "science" in which one may confront the occupational risk—as noted by Weston La Barre—of "feeding multiply contaminated data into...truth machines" (Devereux 1967:viii).

Indeed the "progress" of science is highly dependent on peer recognition and disbursement of funds by the lay converted. It is a grievous oversimplification to *contrast* methodology in science with methodology in the social sciences. The development of either is highly dependent on interpersonal interactions, influence and negotiation of "truth".

Some studies have focused on medicine as a socially constructed field; an interesting collection of essays focused on certain medical events as part of the social, political, economic and cultural processes surrounding such events (Wright and Treacher). Other social constructionists have researched the interactions among researchers in a laboratory (Salk Institute in La Jolla, California) to observe and analyze daily routine work. These investigators demonstrated that, among other things, scientists create order out of disorder, impelled greatly if not exclusively by a thirst for credit and credibility (Latour and Woolgar). Personal goals notwithstanding, scientists do have an ideal about their "mission":

> The myth of science as a purely logical process, constantly reaffirmed in every textbook, article and lecture, has an overwhelming influence on scientists' perception of what they do. Even though scientists are aware of non-logical elements of their work, they tend to suppress them or at least dismiss them as being of little consequence. A major element of the scientific process is thus denied existence or significance (Broad and Wade).

As a result of this, scientific textbooks are nonhistorical, in fact they are often antihistorical. Any reference to the subjective experience is strictly forbidden in scientific literature. Physicist Paul Feyerabend, in his book *Against Method* (Feyerabend), holds that not only are there nonrational elements in the scientific process, but that such elements are dominant, and that success in science depends not only on rational argument but on subterfuge, rhetoric and propaganda.

Medical science and medical practice have been affected by two 20th-century movements, existentialism and phenomenology. Both stress the relationship between the individual and systems, between freedom and choice, with anxiety, truth and belief as aspects of experience. Resonating to such concepts, there is an emergent contemporary disillusion with the incessant pressure built into the reward system to constantly produce "astounding scientific data". There is an emergent demand for a new kind of accountability and a growing awareness of the lack of those virtues that generations of scientists have declared inherent in the scientific endeavor. Few are offended by Kuhn when he sees the transfer of attachment from one paradigm to another as not the sort of battle that can be resolved by proofs, but rather as a conversion experience. As the new fields of medical ethics and malpractice law prosper, confidence has been replaced by mistrust, and a growing literature testifies to the new criticism of scientific research and its protocols. However, the real challenge is to recognize that even though science is not all that some scientists say it is, this does *not* mean that it is *none* of the things they say it is.

THE ART OF MEDICINE IS LINKED TO SCIENCE

Medicine has both cultural and biological dimensions, and therefore it is a kind of applied anthropology. Physicians and medical research scientists, if they are to be effective, must be aware that humans are simultaneously biological and cultural expressions of being. Medicine involves the treatment of patients (the clinical aspect), concern with public research on aggregates (the statistical aspect), as well as with what has been learned from laboratory experimentation—the link with "pure" science. The latter two aspects give us knowledge based on past events (even though they might be challenged and modified in the present), but medical treatment also involves diagnosis, taking personal histories, and gaining confidence for patient compliance, all of which is an ongoing process. Behavior is influenced by expectations about the future; in the doctor/patient relationship this can be good medicine, for in the act of healing the goal to influence outcomes, although the latter is hardly a scientific approach.

When medicine and science join, the epistemic challenge is to combine the several aforementioned aspects of each, in order to understand what and when information is relevant to the current problem (the patient) and to understand that the best one can do is recognize that certainty can only be approximated as a convergence of probabilities (as one diagnoses and prescribes courses of action). Interpretation of pure laboratory data does not address problems that arise from interpretations of clinical trials, which involve such factors as patient selection, patient compliance among other variables that cloud inferences from statistical-epidemiological data.

It should come as no surprise to learn that the courts have their greatest difficulties in resolving cases that involve medical science, and that it is always problematic when medical science has to be represented and interpreted in the courts. In Daubert et al v. Merrell Dow Pharmaceuticals (5), central to the case was the question: what constitutes acceptable scientific evidence, and who should make that decision?

On one level (U.S. Court of Appeals, 9th. Circuit in San Francisco) and at one point in time (1991) it was decided that the stringent standards of peer-reviewed research should outweigh other criteria for determining the admissibility of scientific evidence. In 1993, the Supreme Court decided to give judges more space and power in screening scientific evidence, i.e., they should use "relevance and reliability" rather than merely the peer-review status of the science as the basis for admissibility. Although all of these judicial attempts, from Frye to Daubert, were geared to avoiding the admissibility of "junk science", Daubert clearly directed itself at expanding the scope of what is considered acceptable science to include cutting edge research, which might not relegate itself to a positive peer-review. But that was not the full story, as we shall see later.

WHEN MEDICAL SCIENCE IS A SOCIAL SCIENCE

When the outcomes of medical events become the object of litigation, not infrequently, expertise from the array of "social sciences" is sought by litigants in the dispute. Cultural barriers to understanding produce disputes about: informed consent, non-compliance with medical directives (with ensuing disastrous results), conflicts with the interest of the state in a child's life, questions of professional standards for medical practice, public health issues arising from notions of personal or religious freedom, to provide a few examples. The "insanity defense" and competency to stand trial, particularly in criminal cases, present the need for expertise in psychiatry as well as social science. It may seem odd that experts in fields such as psychology, sociology or psychiatry can be found to testify for opposing views. But even for medical and surgical procedures or pharmaceutical interventions, experts can be located to take opposing views. Leaving aside allegations that experts will say what they are paid to say, it is entirely plausible that most if not all such experts believe their assessments. The point we wish to make here is that opposing assessments are not any more unusual in "hard" scientific medicine than they are in the social sciences.

The methods of social science are easy to criticize: they depend on statistical models, observation (with observer bias), questionnaires (how can you ask a question if you do not already know much the answer?), and on what the observer brings to the enterprise in terms of his or her own cultural baggage (6). The mathematization of social sciences does not resolve the matter of the decision as to which variables are included or excluded. But the most serious criticism is supplied by Devereux, that perception and interpretation of data are distorted, producing counter-transference masquerading as methodology (7). Thus, as noted earlier, "anxiety" about overlap between subject and observer can be detrimental to medical as well as social scientists. As for the law, as Michael Moore noted, "the law is not neutral about the various ways people may be explained" (8). Social sciences can, at least, provide well-reasoned mechanical or statistical models, or decision-making models for human behavior to justify opposing views or beliefs. Medicine, the Art, having incorporated science, is still significantly shaped by cultural values and a subspecialty within it, psychiatry, is most characteristic in this regard. When law and psychiatry engage in dialogue in the court we see the negotiable features of both in bold relief.

In primitive groups, it is the shaman who mediates dystonic episodes, defines deviance and replaces disorder with order. It is a difficult calling to be a healer in a shamanic culture, for the perception of the patient's family that "malpractice" occurred can cost the shaman his life. However, the shaman has several strategies for self-protection; he accepts only the patient he feels can be helped by him, and, more importantly, in shamanic cultures healer and patient share symbolic systems. He is a master of the combinatorial arts, of the "science" and the poetics, and theories of communication accepted by all in the group. Through use of song, dance, drums, costume displays and trance sessions he travels to other worlds for diagnosis and cures.

He is the healer of body as well as the body politic, for group participants in the healing ceremony (Romanucci-Ross 1989).

Lévi-Strauss has suggested that what the shaman is actually accomplishing in the healing ceremony is this: through the symbolism of dance-drama and "traveling to other worlds" he controls mental processes so that physiological processes are altered. His mind-body medicine dissolves ego boundaries as well as boundaries between self and other (Lévi-Strauss).

Shamans in various cultures are quite successful when dealing with cases we would label "psychiatric". In our society, for both doctor and patient, symbolic systems are not shared even though some patients might be closer to the doctor's world than others. In such encounters, effective curing and healing demands recognition that symptom and sign are field dependent and culturally determined. This is especially the case in psychiatry where the "symptom and sign" approach to diagnosis pertains to mood, affect, thoughts, emotion and behavior.

In complex societies it is physicians (including psychiatrists) and men and women of the law who define and restore order in the body and the body politic (9). Psychiatry has the function of establishing that one person's deviance is or is not evidence of a "true" pathology that can be contrasted with deviance which features a calculated intention by a mind that knows the difference between pathologic and normal. When these distinctions are formally recognized for a purpose, e.g. in an investigation or in a court of law, it is called "forensic psychiatry"

Jurisprudence focused on court acceptance of the definition of mental health and on the proper therapy and means of addressing mental dysfunction is also concerned with policies creating relevant legislation and administrative regulations. Much of the data in this area presented in court are of borderline scientific validity. With such caveats in mind, we next turn to the "science of commitment", as it relates to the example of the sexual predator. Many state laws and judicial decisions since 1990, and a Supreme Court decision in 1997, have justified commitment of sexual predators after they have served prison terms on grounds of dangerousness, the potential for recidivism, and problems with treatability. When can an individual be labeled as a "danger to society?"

NOTES

1. William Broad and Nicholas Wade. *Betrayers of the Truth: Fraud and Deceit in the Halls of Science.* New York: Simon and Schuster. 1982 p. 212.
2. Loren Eisely. *The Immense Journey.* New York: Vintage Books. 1946.
3. Horace F. Judson. *The Eighth Day of Creation; The Makers of the Revolution in Biology.* New York: Simon and Schuster. 1979.
4. James D. Watson and Francis Crick. "A Structure for Deoxyribonucleic Acid". *Nature* 1953. 171: pp. 737-738. See also J.D. Watson, *The Double Helix; A Personal Account of the Discovery of the Structure of DNA.* New York: Signet Books 1968.
5. Daubert v. Merrell Dow Pharmaceuticals Inc. 509 U.S. 579 (1993).
6. Lola Romanucci-Ross. *Mead's Other Manus; Phenomenology of the Encounter.* Mass: Bergin and Garvey 1985.
7. George Devereux. *Basic Problems of Ethnopsychiatry.* Chicago: University of Chicago Press, 1980.
8. Michael Moore. Law and Psychiatry: Rethinking the Relationship. Cambridge University press, 1984.
9. Lola Romanucci-Ross. "The Impassioned Knowledge of the Shaman". In: *The Anthropology of Medicine*, op. cit., pp. 215-223.
10. The subsection "to our own…on Values and some research models in Western medicine" contains excerpts (with additions and modifications) from: Romanucci-Ross, Lola and Daniel E. Moerman, "The Extraneous Factor in Western Medicine", reprinted from *Ethos* 16(2), pp. 146-166, ©1988 by American Anthropological Association, by permission of the University of California Press.

CHAPTER 5

THE SCIENCE OF COMMITMENT

The Travails of Forensic Psychiatry

Every culture has a great deal to say about sexual behavior; the concerns are with whom, in what manner, at what times and in which places such behavior can occur. Usually there are specified forbidden relationships and acts as well as tabooed times and places. In this chapter, we discuss a subset of persons in this behavioral area in our culture, those individuals considered and labeled "sexual predators". We will consider the so-called "science of dangerousness", and the role of psychiatry, for the discourse on mental health and illness is highly dependent on cultural values. These are, of course, present in all of medicine but particularly important in psychiatry. "Therapeutic jurisprudence" focuses on how the law influences definitions of mental health by the court as well as appropriate means of addressing mental dysfunction; its proponents hold that policies which shape legislation and regulations should be based on "scientific facts". These facts also form the bases for judicial decisions on both civil and criminal law. We examine here the science of dangerousness as it applies to sexual predator laws in the United States. This is followed by a return to the anthropological in drawing comparisons and contrasts on the "science of commitment" in several other cultures (New Guinea, the South Pacific, Italy and a subculture of our own country—the world of priests, bishops, and the hegemony of the Catholic Church.

SEXUAL PREDATOR LAWS OF THE PAST

The history of these laws reveals an ambivalence in the courts, reflecting public attitudes towards this type of behavior. In some states, laws for "special" involuntary commitment of sex offenders originated in the 1930s, usually following publicity about a sexual offense that was especially disturbing to the community. The justification for loss of liberty through involuntary confinement (as applied generally to the mentally ill, and most recently to sexual predators) has been that it is applied only in circumstances in which the individual has shown an inability to respond rationally to the law's incentive structure (1).

In 1937, Michigan was the first state to enact laws for special commitment of sexual offenders. By the 1960s, over half the states and the District of Columbia had such laws on their books (2). These laws were given various names which reflect the societal perception of deviance and the inability of the offender to maintain control: in some states they were called "sexual psychopath laws", in others "mentally disordered sex offender" acts, and in still others "sexually dangerous persons" acts. They were seen as having two main objectives: first, the treatment of sex offenders with the possibility of seeing results more quickly than would occur under a prison sentence for convicted criminals. By medicalizing sex offenders such laws, in contrast to those which emerged in the 1990s, saw hospitalization as a viable alternative to imprisonment. Secondly, these laws were designed to protect society from the early release of offenders who had served their maximum period of incarceration but had not been effectively cured.

A number of assumptions supported legislation to commit sexual offenders (3). The most prominent of these is that a specific mental disability such as sexual psychopathy responsible for sexual offenses does indeed exist, and that those suffering from this disability are identifiable by mental health professionals. Furthermore, it is assumed that persons so afflicted are more likely to commit serious crimes, especially dangerous sex offenses, than the usual normal criminals. As for predictability of dangerousness, it is assumed that there are adequate diagnostic tools as well as treatments, that mental health professionals have these at their disposal, can predict the dangerousness of these offenders and cure them.

Shortly after these statutes appeared they were targeted for criticism, led by the main objection that treatment for sexual psychopathy was ineffective and that some involuntarily committed offenders would, upon their release, commit a similar offense. So although most of the states provided for indefinite commitment in their statutes, in many instances these offenders were released sooner than they would have been had they been sent to prison. States began to repeal these laws in the 1970s and by 1990 only eleven states and the District of Columbia continued to have statutes for sexual psychopaths. Several groups had recommended the repeal of these laws. Most conspicuous among these were the Group for the Advancement of Psychiatry (1977) (4), a Presidential Commission on Mental Health (1978) (5) and the American Bar Association's Committee on Criminal Justice Mental Health Standards (1984) (6).

In the 1980's there was a strong movement for sentencing reform (7). This movement was fueled by a widening and growing disaffection with the outcomes of rehabilitation, high recidivism rates among paroled offenders, and policy concerns about the disparity of treatments among offenders. States began to abandon indeterminate sentencing and special commitment laws, now favoring a system of determinate sentencing i.e. prescribed or fixed penalties based on the nature of the crime in each case. Under the previous system of indeterminate sentencing, a range of sentences potentially maximizable to the indeterminate sentence e.g. 15 years or more, were available to the court. With such reforms enacted in the 1980s, the sentence for the sexual offender would be for a specific definite period, essentially deleting the discretionary feature of sentencing. No one would be able to win early release or parole because of good behavior or perceived notions of "cure"; alternatively, no one would be detained for a longer period because of perceived dangerousness. "High risk offenders" were singled out in a process that can hardly be called "scientific", and of course some of these (in retrospect) were released "too early". Therefore, determinate sentencing of the 1980's reform could be viewed as having been too rigid.

Problematic aspects of determinate sentencing provided the context, in the 1990's, for reviving interest in indeterminate confinement of sex offenders. But now, policy makers acknowledged a need for a greater concern about community safety and less concern with the treatment feature of commitment (the *quid pro quo* of denial of rights and gifting treatment in return to the offender). Lessons learned from epidemiological research regarding dangerousness, treatability, recidivism, etc. played a pivotal role in justifying this approach, and consequently it took a different turn from the laws of the 1930's. In this new approach, commitment has not been drafted in laws as an alternative to imprisonment. Rather it has become a follow-up after the offender has served his "time", a feature was viewed as providing a preventive detention (8). Such detention, however, flies in the face of legal postulates (deeply ingrained in our Bill of Rights) affirming that harm has to have occurred before restraining the harm doer. Exceptions, as pointed out by some, could be considered in situations in which preventive detention outweighs the cost to liberty (9). Justice O'Connor in her concurring opinion in *Foucha v. Louisiana* essentially endorses a balance test, i.e. a mental impairment requirement for civil commitment as against the government's interest in detention (10).

The history of the laws regarding mentally disordered sexual offenders indicates that our society is ambivalent about people who commit acts of sexual violence (11). On the one hand, these offenders are seen as clearly wicked, dangerous and deserving of substantial penalties. However, there is also the general sense that there must be "something wrong" with sexual deviants, for many cannot comprehend the reasons for deviant sexual behavior and therefore take the position that such individuals are mentally disturbed. Furthermore, people appear to be more frightened by sexual offenders than by people who commit other violent offenses. Do sexual offenses strike closer to our sense of personal identity, or a violation of the self and of intimacy or the potential for intimacy? It is interesting to note that legislatures have never adopted commitment laws for arsonists, and armed robbers, to name a few socially repugnant acts.

SEXUAL PREDATOR COMMITMENT LAWS: THE 1990s

The first of the new Commitment statutes of the 90s was Washington's Sexually Violent Predator Statute (12), adopted shortly thereafter by seven other states (13), and then by two additional states. Many other states, such as Pennsylvania and New York, are currently reviewing their laws on sexual psychopathy with a view to changing them in conformity with the Washington law. Almost all states provide for post-prison commitment of offenders, while Illinois allows for confinement and treatment of sex offenders only as an alternative to criminal prosecution. Every state that has passed this law, except California, has specified that commitment may be for an indefinite duration. In California, the commitment is for two years, after which an additional petition and jury trial are necessary if the state believes further confinement is necessary.

The Washington statute, initiating the new movement for indefinite confinement, resulted from the case of Earl K. Shriner. Shriner completed a 10-year sentence for kidnapping and assaulting two teenage girls. Two years later on May 20, 1989, he raped a seven-year-old boy and cut off his penis. A great public outcry impelled the Governor of the state to appoint a task force to study the problem of sexual offenders and to make recommendations. The task force produced a report which led to the Community Protection Act in the state (14). Among the features of this act were requirements that sex offenders, when released, register with the police in the communities where they live (often referred to as Megan's Law), post-release supervision for certain sex offenders, increased penalties by 50% on average for sex crimes, and a new law that allowed for the civil commitment of persons deemed to be "sexually violent predators".

Unlike the 1930s legislation, involuntary commitment was not intended now as a substitute for punishment through the criminal system. It was expected, rather, that the offender would serve full prison term first, then be evaluated for involuntary civil commitment, which could by law be indefinite. It has not been necessary to show that the offender so confined has committed a recent sexually violent act. The state's sole duty has been to show that the offender is a "sexually violent predator", which in Washington and Kansas (15) is defined as a person "who has been convicted of or charged with a crime of sexual violence and who suffers from a 'mental abnormality' or personality disorder which renders the person likely to engage [in the future] in predatory acts of sexual violence".

Both the Washington and Kansas statutes specify "mental abnormality"; when examined closely, the statutes reveal that the underlying intent of the legislation is essentially to confine virtually all sexual offenders (not just violent sexual predators) for unlimited periods of time to protect the community. According to the Washington law, "mental abnormality" refers to "a congenital or acquired condition affecting the emotional or volitional capacity which disposes the person to the commission of criminal sexual acts". The second part of the definition, "personality disorder", is not defined in the statute. Involuntary commitment for the sexual predator is premised on the proclivity to commit future harm and not on current blameworthiness.

DEFINITIONS OF MENTAL ABNORMALITY

"Mental abnormality" as defined above does not comport to recognized diagnostic criteria described in the primary reference for mental illnesses, the American Psychiatric Association's Diagnostic and Statistical Manual of Mental Disorders (DSM-IV) (16). Legislation can and does create its own terms: it is under no requirement to comport to criteria established by disciplines such as psychiatry. The critical issue in legislative drafting legislation is not whether a statutory definition comports with definitions provided by a science, but whether it makes sense on its own bases (17). In this legislation "mental abnormality" refers to a psychological state in which a biological or environmental condition affects one's emotional or volitional capacities such that it predisposes an individual to engage in criminal sexual acts (18). This definition incorporates the causation of every conceivable behavior, since biological and environmental conditions encompass the universe of factors which cause human conduct. There is nothing about the definition that narrows the class of persons that can be designated as "abnormal". The definition is entirely predicated on the presence of "criminal sexual acts" (19) and becomes a tautology, i.e. "mental abnormality" equals "criminal sexual acts" and vice versa, an equation that provides justification for commitment. "Mental abnormality" is broad and elastic, including nearly any deficit, symptom or historical fact regarding an individual who engaged in "criminal sexual act(s)".

Essentially, the underlying theory for these new commitment laws seems to rest implicitly on the perception that those committing criminal sexual acts lack the ability to control their deviant behavior (20). This basis was highlighted, as we will see shortly, in the case of *Kansas v. Hendricks* (21) which was decided by the U.S. Supreme Court in 1997.

U.S. SUPREME COURT IN HENDRICKS

Statutes providing for the commitment of sexual predators were challenged in many states that passed this legislation (22). The Kansas statute was challenged in 1996 in the case of *In re Care and Treatment of Hendricks* (23). Leroy Hendricks, now in his mid-sixties, had a long history of pedophilic behavior, having molested children repeatedly throughout most of his adult life. In 1955, Hendricks exposed himself to two young girls and pleaded guilty to indecent exposure. Two years later, he was apprehended while playing strip poker with a teenage girl and was convicted of lewdness. In 1960, he molested two young boys at a carnival and was imprisoned for three years. Shortly after being released, he molested a seven-year-old girl and was arrested and convicted. Subsequently arrested and imprisoned in 1967 for sexually molesting a young girl and boy, Hendricks was finally imprisoned for ten years in 1984 for molesting his stepdaughter and stepson. After his release, he was examined by a state physician, under the Kansas Violent Predator Act, and was found to be an uncured pedophile. A Kansas jury determined that he was a "sexually violent predator" and committed him (involuntarily) to a psychiatric hospital.

Hendrick's commitment was challenged under the theory that it violated the Substantive Due Process Clause of the 14th Amendment of the U.S. Constitution. The Kansas Supreme Court agreed, noting that the Kansas Violent Predator Act permitted the confinement of persons who were dangerous but not "mentally ill", as there was "no support in the Act that a finding of mental illness is required". The court referred to the 1992 U.S. Supreme Court's holding in *Foucha v. Louisiana* (24) which held that commitment to a mental institution requires that the state prove by "clear and convincing evidence" that the individual is both mentally ill and dangerous, and that dangerousness in the absence of mental illness does not justify commitment. Furthermore, continued involuntary detention of an insanity acquittee who no longer suffers from mental illness on the basis that he has not shown he is dangerous to himself or others violates due process. Hence, the Kansas Supreme Court claimed that "mental abnormality" and "personality disorder" (the latter undefined in the statute) do not constitute mental illness, thereby taking the position that the term "mentally ill" designates a precise medical condition (perhaps in accordance with the DSM IV; this is certainly in conformance with Kansas' Treatment Act for Mentally Ill Persons--K.S.A. 59-2901 et seq. which identifies "mental illness" as required for involuntary treatment). Therefore, civil commitment statutes that do not contain the term "mental illness", according to the Kansas Supreme Court, cannot withstand challenges based on due process arguments (25). The Court further held that the segregation of sexual violent predators from the public was the overriding concern of the legislature in drafting the Kansas Sexual Predator Act.

The Hendricks case reached the U.S. Supreme Court which, in 1997 (26), declared that the Kansas' Sexually Violent Predator Act was constitutional, rejecting Hendrick's fundamental challenges. Justice Clarence Thomas holding for the majority dismissed the idea that the mental condition in question must be officially recognized or comport to the nomenclature of diseases designated by the mental health community. He stated the Supreme Court has "traditionally left to legislators the task of defining terms of a medical nature that have legal significance" (27). Furthermore, he dismissed the arguments of double jeopardy, claiming that the commitment was not punitive since it does not involve either retribution or deterrence; it is not retributive because no criminal responsibility is implied. On the issue of deterrence he referred to the fact that persons committed under the act have a "mental abnormality" or "personality disorder" which prevents them from exercising control over their behavior. The Kansas Supreme Court also had found that sexually violent predators are not responsive to treatment. (In previous laws on the commitment of sexual psychopaths, treatability was a prerequisite (28).) The majority holding of the U.S. Supreme Court recognized the problems with treatment of offenders, but stated that the Constitution does not prevent a state from detaining those who pose a danger to others and for whom no treatment may be available.

- In effect, the U.S. Supreme Court in the Hendricks case distinguished its position from *Foucha v. Louisiana* (and the preceding body of cases from *O'Connor v. Donaldson* through *Addington v. Texas*) (29). The critical distinction from prior cases is that Terry Foucha, who had a personality disorder, was not prevented by virtue of that mental condition from controlling his conduct (30). Hendricks had admitted at the time of his commitment that he still maintained sexual desires for children and that these desires could not be controlled when he was under stress. Therefore, these cases exemplify the position that, when uncontrollable by the perpetrator, individual dangerousness requires commitment; when controllable it leads to resolution through the criminal process, i.e. imprisonment. In Kansas v. Crane (decided January 22, 2002) Justice Breyer, writing for the 7/2 majority indicated that the state's belief in an offender's future dangerousness could not be the defining standard, but rather the state must believe that the individual shows "serious difficulty in controlling behavior". Thus, the commitment criteria have to include a finding that the offender is completely unable to control his violent sexual proclivities and therefore poses a danger to the community. This is a stricter standard that that which the district court applied in allowing Crane to be committed. Therefore the U.S. Supreme Court reversed the Kansas Supreme Court in this case (*Kansas v. Crane*: argued October 30, 2001, decided January 22, 2002, 269 Kan. 578 P. 3d. 285, vacated and remanded 2002) (31).

In the 1980s even before notorious cases, such as the Hinckley case (32), which led "reform" away from using the volitional aspect of the legal insanity test, there was a strong movement to limit the insanity defense to individuals with cognitive impairment, i.e. lacking the capacity to know the nature and quality and moral correctness of the act. Prior to these changes, however, some jurisdictions once permitted a defense of insanity predicated on the presence of an "irresistible impulse". This test focused on the offender's volitional capacity, i.e. even in the presence of cognitive understanding of the nature and wrongfulness of an act, an insanity acquittal would be allowed if the defendant could establish that his mental disorder compelled him/her to commit the offense, that he/she had an irresistible and uncontrollable impulse to commit the act (33).

The distinction between impulses not resisted and irresistible impulses was difficult, and the establishment of one over the other in individual cases so problematic that both the American Psychiatric Association in 1982 (34) and the American Bar Association recommended elimination of the "control" test for legal insanity. As for pedophilia and sexual predator offenses, there is no empirical evidence, nor for that matter theoretical support, for the notion that such offenders are unable to prevent themselves from acting on their sexual impulses. Furthermore, the Kansas program (as well as most of the other states) delays civil commitment for the sexual predator until after the offender's prison term has been completed. Even though the statute doesn't require that the individual be considered treatable before he is adjudicated a sexual predator, from the perspective of therapeutic jurisprudence the system presents serious antitherapeutic effects. While in prison, these sexual predators are essentially receiving no treatment for their "mental abnormality". Treatment delivered closer to the time of the offense is more likely to have some effect on the transgressor than treatment delivered years later.

We have seen major swings of the pendulum during the 30s through the 90s in the treatment of sexual offenders, as well as the mentally ill at large. Initially, the operative thesis was that afflicted persons lacked competency to control their behavior and therefore the state had to act as *parens patriae* by committing and treating these people for their own good. During the Civil Rights movements of the 1960s, for mental patients generally, the move was away from indeterminate commitment and *parens patriae* towards emphasizing greater autonomy for the mentally ill. For sexual offenders, the switch came in the 70s and 80s due to disaffection with the results of treatment. The pendulum then shifted to use of the criminal law over medicalization, but it moved once again with the new legal changes in the 90s. Commitment for sexual predators (as well as notification laws which will be discussed briefly at the end) reinstated the role of the state as guardian, as *parens patriae*.

This shift back to commitment revived some of the old reasons for justifying loss of liberty. The statutes state that a sexually violent predator must be considered "likely to engage in predatory acts of sexual violence". This requirement relies on the ability of professionals to predict an individual's future sexual conduct.

DANGEROUSNESS AND RECIDIVISM: WHAT OF PREDICTION?

Within the past twenty years a rich and extensive literature on studies concerned with the prediction of dangerousness (35) have indicated that, for the most part, clinicians do not reliably predict violence (36). In fact, many of these studies were introduced in landmark U.S. Supreme Court cases in the late 70s and early 80s to support this proposition (37). Recent research seems to suggest that the mentally ill (contrary to many earlier studies) are "somewhat more likely" to be violent than the general population (38), but the inaccuracy of the clinician in assessing and predicting dangerousness persists.

In 1993, a methodologically persuasive study demonstrated that clinician predictions of violence had a 60% sensitivity (true positive proportion) versus an expected chance accuracy rate of 50% (38). Indicating that clinicians are "relatively inaccurate predictors" of violence, the authors also noted that predictions for male patients exceeded chance by a small margin, and for female patients predictions were no better than chance. The study involved a sample that was particularly violent. Predictions of violence decreased in a random selection of emergency room patients who exhibited a "slight" to "no" degree of violence proneness. One of the more provocative conclusions of this study is that clinical judgment is not necessarily more accurate than reliance on a history of violence as the basis for predicting future dangerousness.

In 1994, a study of 58 reanalyzed data sets from 44 published studies on the prediction of dangerousness (39) suggested that past behavior is the best basis for predicting future violence, that clinicians predicting dangerousness generally exceed chance but inevitably make errors and that non-clinicians provided with information about past behavior can outperform clinicians who rely completely on data obtained from a clinical interview. Some researchers extol actuarial (40) over clinical predictions of dangerousness (41). The actuarial method relies on statistical factors of a group as it provides a determination of the likelihood (rate or percentage of an aggregate number at risk of a particular behavior) of future dangerousness, contrasting with clinical assessments which aim to establish prediction of dangerousness in particular individuals.

On an examination of a high-risk sample (42), it was concluded that there is no highly accurate method of predicting violence, and that little progress has been made in recent years on risk assessment for dangerousness of mental patients. It was further determined that even risk assessors who were considered to be best at this evaluation provided predictions that were of little value. Applying several socio-demographic variables resulted in better results for prediction than either clinical or actuarial assessments.

SEXUAL OFFENDER RECIDIVISM

The recent generation of sex offender laws was promoted by advocates who argue that new expertise has emerged which allows for a firm foundation for such laws (43). The basic premise has been that experts now have the means to predict sexual recidivism. Hanson, examining whether there is firm empirical support for this position, reviewed risk assessment methods and what is known about recidivism rates for sex offenders. He concluded that actuarial risk scales grounded on static factors such as number of prior sexual offenses and deviant sexual preferences--preferences that are most unlikely to change--demonstrate promise for identifying relatively accurately which sex offenders have a basic tendency to sexually reoffend (44). He concedes that mistakes on predictions are inevitable; nonetheless, he considers that sufficient expertise exists for predictive-preventive strategies to confine sex offenders who are at high risk of re-engaging in deviant sexual acts.

A relatively recent meta-analysis of treatment for sexual offenders discovered that the sexual recidivism rate was approximately 19% for treated offenders and 27% for the untreated (45). Moreover, most recidivism even by violent offenders is generally nonviolent in nature, and, in a review of the likelihood of a recidivism being violent, it was found that subsequent arrests involved trivial rather than serious offenses, in a ratio of three to one (46). Many researchers believe the actuarial prediction for determining which sex offenders are at most risk of reoffending is superior to clinical assessment as a predictive method, if one knows the sex of the victim and his/her relationship to the offender, the offender's history, age of arrest, and marital status (47).

Several observations come out of this body of research on sexual offenders. First, most recidivism, even by offenders who committed violent acts in the past, is non-violent. Highly accurate prediction of future sexual recidivism as with the successful prediction of dangerousness is currently beyond our reach. In addition, actuarial methods for prediction are superior to clinical assessments; and inference from the studies suggests that sexual offenders are not more likely than other violent offenders to recidivate.

Predictors of recidivism take on more substance when sexual offenders are differentiated into specific groups. It has been shown that it is necessary to follow sex offenders for long periods of time, sometimes over a 15 or more year period (48). One study showed that approximately 10% of a sample of 197 child molesters were reconvicted more than ten years post release from prison; some convictions actually occurred after more than 30 years post-release (49). Also, there are differences in the types of recidivism, e.g. rapists are just as likely to be convicted of a nonsexual violent offense as a sexual one (50).

- The recidivism--sexual and violent--of a group of 54 rapists had been assessed in a forensic hospital and released after serving a prison sentence. The researchers used predictors such as age, history of offenses, psychopathology, marital status and phallometric deviance scores (51). The latter is standard used to differentiate rapists from non-offenders by means of phallometric assessments (measurement of penile tumescence during the presentation of a variety of sexual stimuli) (51). Despite the short follow-up period (46 months) they found that the violent recidivism rate was high (43%) and 28% were convicted of new sexual offenses. A meta-analysis of studies available by the end of 1995 examining predictors of sexual recidivism reviewed 87 articles involving 61 data sets. The study included 28,972 sexual offenders with a median follow-up period for the studies of 4 years and a median size of 198.

The strongest predictor of sexual recidivism was deviant sexual preferences as measured phallometrically (52). The next most important variables were early onset of sexual offending and prior sexual offenses, followed by age, and marital status (never married).

Studies of recidivism among sexual offenders are somewhat more optimistic than those of prediction of dangerousness among general violent offenders. There are some groups of offenders--rapists, child molesters--where prediction based on a set of variables including phallometry, and past history of sexual and violent offenses is significantly better than chance, although many argue that clinical assessments should be replaced by the latter (53). The issue of accuracy becomes critical in the case of Sexual Predator Laws where the consequences of commitment attend following imprisonment and may be for indeterminate periods of time (54).

DOES TREATMENT ALTER RECIDIVISM?

There are few controlled studies on the benefits of treatment for sexual offenders. Existing treatments consist of non-behavioral psychotherapy, somatic treatment, surgery, drugs, behavior and cognitive-behavior therapy and combinations of these (55). Psychotherapy was the most prevalent type of treatment until the past 15 years. Outcome studies have been conducted (psychotherapy was the most prevalent treatment until 15 years ago) and have been discouraging. There seems to be little evidence that these treatments reduce violence and recidivism in various study populations (rapists, child molesters) (56). In some cases, psychotherapy may in fact increase the likelihood of new sexual offenses (57).

Surgery for sexual offenses primarily involves castration. It is very rare in the United States where it has been unpopular primarily for ethical reasons, but there has been used in Europe. Many studies of the effects of castration involved men who were not sex offenders. A Swiss study involved 121 male sex offenders who were castrated (50 refused the procedure). In a follow-up of five to thirty years, 7% of the castrated versus 52% of the uncastrated committed subsequent sexual offenses (58) suggesting castration may reduce sexual desires and risk of re-offending. It is important, nevertheless, to recognize that rapists commit a lot of nonsexual violence.

Two are popular treatments, cyproterone acetate and medroxyprogesterone, present fewer ethical dilemmas as the effects are essentially reversible and treatment outcomes have shown relatively good results. In one study of paraphilias, over half of whom were pedophiles, the treated group followed over five years had a rate of relapse of 15% whereas the untreated group's rate was 68% (59). The problem has been primarily one of compliance. Few offenders voluntarily accept medications to decrease testosterone, and those that do are unlikely to remain on the regimen for an extended time. Other drugs have been used increasingly--particularly the serotonergic drugs such as Prozac or Zoloft and some preliminary studies report beneficial results.

Behavior and cognitive-behavior therapy has been problematic. In laboratory context a group of individuals can be treated through conditioning, measured arousal and these have shown beneficial results. The problem is whether this approach results in benefits to these individuals when they are re-introduced into the community. This has yet to be shown in any significant studies. There do seem to be benefits from the use of group therapy based on the Alcoholic Anonymous model, i.e., a support group resulting in breakdown of individual denial and rationalization as well as useful mechanisms for learning to avoid provocative circumstances.

CULTURAL ASIDES ON EMPOWERING THE OFFENDER/PATIENT

This first case presented above featured predominantly "soft" criteria in application of the scientific method. Let us refer again to the most fundamental bases of science: accurate observations that allow reliable prediction coupled with testing of hypotheses that can define cause and effect relations. "Soft" as they may appear, the scientific approaches in the above case are not so wide of the mark. We have seen that investigators in the "the science of commitment" have included psychometric and sociometric approaches, and even a physiological measurement (penile tumescence) which can be considered a "hard" scientific criterion. But there are still other important considerations in the behavioral sciences. Thus, in the "science of commitment", not only the state of medical knowledge concerning mental health but also socio-cultural-political factors, such as fairness to the offender and the protection of society must affect application of the law. Our society has made choices based on certain premises about individual rights and the role of society to protect its constituents, particularly its children. We have made certain assertions about the certainty of knowledge about behavior (the courts do define "intentionality" in specific circumstances). Such moments of assuredness brought decisions as to whether an offender went to prison or to a treatment center, or was given a prison-plus-treatment sentence, either fixed or adjustable. When should the state be *parens patriae*? We find attitudes shifting in the 1930's, the 1960's, and the 1990's. These changes can be linked to political attitudes, to parties in power and the political strengths of their constituencies. For example, the movement to deinstitutionalize mental patients in the early sixties was fueled as much the Civil Rights movement in the United States as by the advent of psychotropic medications that allowed many patients to function outside the hospital in community centers or halfway houses. The granting of rights for persons formerly regarded as an underclass was now extended to include mental patients. Unfortunately, for many, the new abode for this freedom of choice became the streets, or status as "guest" in a prison.

The core argument about "volition" or "free will" and "intentionality" has occupied the court heavily in deciding what to do with a sexual predator; it also marks a core difference between our two major political parties in stating "self evident" truths. One party holds that the individual should be a able to "make it" on his or her own, regardless of circumstances (going so far as to suggest that the circumstances are proof of lack of ambition and any desire to improve one's lot); the other party takes the adversarial position that the state has an obligation and duty to help in every way and at whatever cost those who are victims of circumstance and in need of help. Despite these differences there is at least one concrete historical moment in which both parties opted for the same solution to a social problem: doing away with the institutionalization of the mentally ill. The more liberal party reasoned that this approach would be helping mental patients recover and learn how to be self supporting, while the more conservative party argued that the movement would save money for taxpayers which then could be used to benefit former mental patients by providing more opportunities in the private sector to change the social scene.

FROM THE SHORES OF THE ADRIATIC TO THE SOUTH SEAS

Another path toward deinstitutionalization was exemplified, in Italy, by what has become known as the Basaglia movement or existential psychiatry an attempt to translate a philosophical position into praxis. In the 1960's and early 1970's it appeared very attractive to a number of psychiatrists and some politicians. As in the United States, the problem of deteriorating buildings that housed mental patients, the escalating costs of care and the advent of new and effective medications, coupled with a changing political scene, provided the impetus for deinstitutionalizing. Dr. Franco Basaglia and a group of like-minded professionals were instrumental in the passage in 1978 of Law 180 which attacked the concept of "dangerousness" (to self and others). It demolished the concept that had allowed involuntary commitment, gave priority to psychiatric treatment, provided for community based services and stipulated that all admissions to asylums had to end after 1982. The Basaglia movement itself came to an end in the mid 1980's; the overtly expressed reason for its termination was that sources of local and government funding had disappeared because of a recession. A review of the many studies of this social experiment i.e. allowing all patients to leave the *manicomi* (mental hospitals) and re-enter society in various niches, did not provide the success story that many had hoped for. Its demise was followed by a plethora of controlled studies by researchers on the mental reform in Italy (Burti et al, Calabrese et al, Crepet 1990, Tansella et al, Lovell and Scheper-Hughes). Even though a number were successfully reintegrated with family and friends and appeared to benefit from medications and occasional visits to mental health professionals, no one disagrees that a number of patients did indeed prove harmful to themselves and to others. In Italy, most constituents of political parties were not interested in numbers of successes and failures. One injured child was one too many. Italian culture is also characterized by an overriding concern for the child, rendering unseemly any discourse about the "rights" of anyone who might put the life, integrity, and future of a child at risk (Romanucci-Ross 1991 pp. 139-147). This is not to imply that Americans are not concerned about every child's safety, for we are. But we keep returning to other issues such as civil rights for those judged "dangerous", rights of defendants in criminal trials, legal challenges resulting in accretion of laws, and so forth, advanced by established movements encompassing other concerns. Fearful that one group might be singled out as object for discrimination, our medical model for "dangerousness" does not dwell, as did Basaglia's, on the very definition of madness. We did, however, ask some of the same questions: Who is dangerous and therefore to be hospitalized? What are the cost/benefits of the legal decisions, both in dollars and in sound social policy? What of the right to treatment and right to refuse treatment? However, in Italy, when the debate, both verbal and "acted out", came to a close, many felt that surely some are falsely imprisoned, and that, yes, madness is a negotiable state, but what is non- negotiable is the absolute safety of the child (Romanucci-Ross 1996). For another more pronounced and absolutist approach we return to the Admiralty Islands of Melanesia (see footnotes 1 and 2 of Chapter One).

Anthropological field research is an experience in abstraction. We note and analyze particulars, and put them into brackets as we search for universals in behavior and in thought systems, in an attempt to isolate deep structural similarities and differences in all the cultures of the world. Beyond surface differences we try to find similarities in structures, even as we seek to write in what is referred to as "thick" description. And so we recall the case of Hitawari, of one of the Admiralty Islands villages, to illustrate how, in a primitive society, what we have been concerned with above in the commitment controversies is considered, expedited, and resolved.

Hitawari was in the medical line-up in her village while the anthropologist, who had just arrived that week was dispensing help to villagers with problems that could be accommodated in such a setting or was making referral recommendations (which were rarely acted upon by the person who needed the help) (Romanucci-Ross 1985). Hitawari was unusual particularly because she consciously presented a psychiatric problem, describing herself as agitated, given to states of depression and incessant weeping, harboring death wishes and being subject to bouts of vomiting. Her husband had thrown her out of their hut and had kept the children: he also had a young girlfriend, who lived in the same small village. Villagers who saw Hitawari in the line-up rushed to tell the anthropologist (Lola Romanucci-Ross) that Hitawari was crazy (*long-long* in Pidgin English), and shortly after this all the adults of the village conducted a sort of moving symposium (during the twilight hours after a day of fishing at sea or returning from hunting in the inland forest). The points they explored with each other were then given to the anthropologist, who had not solicited them. The chief reported that he had examined the brains of dogs and *kapul* (opossum, which was used as food) and had come to the conclusion that when all the "ropes" in these examined brains had become *bagarap* ("buggered up" or tangled), then the person would lose any trace of rationality and become *long-long* (crazy). The anthropologist labeled this opinion in her notes as the biomedical model of madness. He then noted, as had others of his group, that when the American Navy had a base near their particular island, they saw an unimaginable number of people. This population density surely would cause individuals to become crazy, for how could you keep all the rules for proper behavior in your head at all times under such circumstances? We must have a large number of crazies in America, right? At least the most fragile among the anthropologist's people (Americans) would lose control under the conditions that we might call stressful, he suspected. (In her notebook this became the socio-cultural stress model of madness). There were, in addition, a noted list of Hitawari's actions that clearly were not sane: interrupting the anthropologist as she was eating dinner to bring a child with an earache that it might be cured, that she put flowers on the table though flowers are not eaten, talked to dogs though dogs do not respond, prayed all night in an abandoned church and fretted about her husband's affair. Although the anthropologist assured them that Europeans and Americans put flowers on the table but do not eat them, talk to dogs but do not expect rejoinders, suffer over affairs of spouses, etc. they are not considered mad for this. The chief's response was that Hitawari is not an American nor a European, and therefore she would not be doing this in the village if she were sane.

When asked what they would have done with such a person in the old days, they said that she would have been killed, or ejected from the group in some other way. After some deliberation it was decided by villagers that they would have the two women fight until one of them "drew blood". And so they did. After a furious match, Hitawari bit off a finger of the young woman and they both were transported, by the anthropologist's outrigger canoe and crew, to the Australian Naval Hospital, where a psychiatrist had Hitawari placed in a hospital - for-all-reasons (mind or body) in New Britain. Before she left, she said to the anthropologist, "They say I am crazy, but I am not, for if I were, I would be running around the village and on the beach with no clothes on. Now that is crazy". Nevertheless, as the chief had indicated, there are many ways to eject the mentally disturbed, using all means at the village's disposal (even the presence of white men and women whose "work" is a mystery to them). As usual, "community protection" won the day at a high cost for Hitawari.

It would be an error to see all this as oversimplified and perhaps even cruel. What needs to be appreciated in this abbreviated vignette is this: Rich though it was in mythology, art, poetics, and structured interpersonal relations, this small isolated group had a material culture that was not complex (i.e. "primitive"). Life on a very small island in the South Pacific was precarious and therefore the group had very strict rules of conduct, for group survival depended on trust and cooperation among all members. Most importantly children were considered a precious resource to be protected by everyone under all circumstances. Dissonance on any level or for any reason had to be eradicated or contained. In such a culture the individual does not have "rights" as such; personhood is only a temporal physical expression of filling a slot in a kinship group, which can and will be filled by another person in time... a "time-share" identity for the individual. But the social structure has always included the animal and plant world linked through one's totem and totemic ancestors. Perception of connectivity implies that perturbation in any part of the system resonates throughout all the systems. Therefore causality for anything is sought in any and all systems, and discord is never taken lightly. There is neither room nor time to "work things out" for the person. So, there were no regrets by anyone in the village, not even her immediate family, when Hitawari was sent off to another distant island (in a distinctly different culture), effectively ejecting her from all social systems that she had known. Such a moment, facilitated by the "white man's" medicine that labeled Hitawari a patient and sent her to a hospital in a foreign distant island, was savored and appreciated by the entire group.

No need to belabor the obvious and observable surface structure differences between Hitawari's culture and a complex industrialized culture. In our own culture, we recognize that some individuals compromise group harmony and we too have ejection rules though they are more "informed" and expressed in a more recondite fashion. We can afford the luxury of negotiability of these concepts. We also have systems within systems (sets and subsets of social groupings), some having a legacy of rules of conduct which are not always compatible with those of the larger dominant social group. These differences may go unnoticed or cause little concern until the sequelae suddenly become unacceptable, causing perturbations leading to revisionary views about mores and laws in both systems. We consider briefly the following current example:

RELUCTANT BISHOPS AND THE SEXUAL PREDATOR

In the year 2002 in America there was to be found an accumulation of allegations from young and middle aged men who referred to themselves as survivors of sexual abuse at the hands of their priests when they were about 12 to 14 years of age. Such revelations, "incredible" in and of themselves, were followed by dramatic displays of a new twist on sexual predation and of social forces shaping perception, all of which was widely reported by the press and other media. (L.A. Times June 17, 2002). Many media discussions focused on the excessive number of homosexuals in the priesthood, the outmoded ideal of celibacy, or the "time-has-come" argument for the ordaining of women as priests. Proponents of such arguments saw them as causal, but there is no data to indicate that this is so. To the contrary, we can point to a large number of incidents and cases indicating that many sexual predators do not appear to be homosexual, nor are they unmarried, and that women in the workplace do not stay the hand of the sexual abuser.

The events were most painful to the Catholic laity, who soon noted that the Bishops whose diocese had such incidents most often resolved the problem by transferring the offending priest to another parish, where he often continued the practice in another venue. Many Catholics, inferring that certain Bishops cared much about reputation and very little for the young lives that were severely damaged, accused the Bishops of being virtual accomplices and demanded they resign or be ejected by Edict. Soon one was able to observe agreement among conservatives and liberals in the laity concerning punishment for the Bishops, but not for the same reasons. Conservatives placed the blame on the acceptance of homosexuality in ordained priests by church, while the liberals blamed the refusal of the church to allow priests to marry and refusal to accept women in the ministry. They also wanted Bishops to agree to "zero-tolerance" for such sexual abuse.

Bishops rejected the zero-tolerance policy, but after a frontal attack by many in the laity (and relentless media coverage) they agreed with the concept, adding that such priests would no longer be able to celebrate mass nor participate in any formal church function. They also later addressed their own accountability in the matter. In a formal document they also determined that such acts should henceforth be reported to the local appropriate authorities (L.A. Times June 22), (It is doubtful that this admirable document will be given the same exposure that the previous scandal that provoked it received). The meeting that produced the document was almost contemporaneous with several Grand Jury recommendations that included the importance of instituting laws that would make it a felony to knowingly send a sexual predator priest to another parish; also, priests who had committed such an act, or reported to have committed it, had to be turned into the civil authorities. Transcendental, hegemonic unto itself and inassimilable, the Catholic church in America recruits priests from a pool of (mostly) American men, who bring American culture with them into the new role. Historically the church has always addressed internal problems within its own hierarchy and with its own rules; these new rules of conduct proposed by both the Bishops and several grand juries almost contemporaneously, if implemented, this will be a watershed in culture history.

OUR CULTURE, OUR PROBLEMS, OUR DATA

As we considered sexual predators in American culture we found an illustration of how frequently inaccurate or imprecise data can be used to bring about legislative changes and judicial decisions that can significantly truncate individual rights. We have seen that the analytical and empirical data on the prediction of sexual offenses suffer from many of the same conceptual difficulties that are observed in predictions of general dangerousness.

Therapeutic jurisprudence was designed to produce law reform that will enhance the law's potential as a health promoting force (60). From the point of view of therapeutic jurisprudence, the sexual predator commitment laws (and those involving notification--see endnote #1) are distinctly anti-therapeutic. The decision in the Hendricks case does not require treatability and, in fact, appears to reject the contention that treatability should be a prerequisite for civil commitment. This fact alone creates a strong presumption that current laws are merely intended for community protection (not for therapy of offenders deemed "abnormal"). Commitment is essentially a prescribed program of "preventive detention" which should be unconstitutional, but is cured of that "unconstitutionality" by attaching the program to mental status issues. We are left with an illustration of the use of scientific attempts seeking cause and effect (with variant degrees of validity) for the purpose of promoting or prohibiting social policy agendas.

NOTES

1. This discussion on Sexual Predator Laws is essentially limited to the issue of involuntary commitment. However, the Washington Statute also provided for community notification which results in major intrusion into individual privacy. These laws have come to be known as "Megan Laws" because of a case in New Jersey and may require notification for 10 years or a lifetime depending on the statute. All states (and recently a federal statute--Jacob Wetterling Crimes Against Children and Sexual Violent Offenses Registration Program, 1995) require registration by sex offenders in community and the provision of information to citizens. Statutes vary in what they require from basic demographic information (age, name, social security number, race, sex, hair and eye color, height, weight, addresses, description of past offenses etc.) to fingerprints, blood samples, etc. In some states, this information is kept in files for the community. In others, it may be placed on the Internet, in the local newspaper. In still others, the offender once released may have to send notice to neighbors within a one-mile radius. Some states even allow for affixing a bumper sticker to the offender's car, placing signs on his residence, or labels on clothing.
2. See Alan H. Swanson, Sexual Psychopath Statutes: Summary and Analysis, 5 *J. Crim. L & Crim.* 215 (1960).
3. *ABA, Criminal Justice Mental Health Standards*, Commentary to Standard 7-8.1, at 455-457 (1989).
4. Group for the Advancement of Psychiatry, *Psychiatry and Sex Psychopath Legislation: The 30's to the 80's* (1977).
5. 4 Task Panel Reports, Submitted to the President's Commission On Mental Health 1978, at 1461 (1978).
6. See footnote #2.
7. See Michael H. Tonry, Real Offense Sentencing: The Model Sentencing and Correction Act, 72 *J Crim. L. Criminology* 1550, 1551 (1981). Also, Von Hirsch & Haurahan, Determinate Penalty Systems in America: An Overview, 27 *Crime Delinq.* 289 (1981).
8. Stephen J. Morse: Symposium: Blame and Danger: An Essay on Preventive Detention. 76 B.U.L. Rev. 113 (February/April 1996).
9. Id.
10. *Foucha v. Louisiana*, 504 US. 71 (1992).
11. See American Psychiatric Association's, *Task Force Report On Sexually Dangerous Offenders*, to be released Spring 1999.
12. *Wash. Rev. Code* Section 71.09.010 (Supp. 1990-1991).
13. Arizona (*Ariz. Rev. Stat. Ann.* sec. 13-4601 et seq. (Supp. 1996-1997); California (*Cal. Welf. & Inst. Code Ann.* sec. 6600 et seq. (West Supp. 1997); Illinois (1938, Sexually Dangerous Person Act, Chapter 116 see 1997, Sexually Violent Person Act, Senate Bill 6); Kansas (*Kan. Stat. Ann.* sec. 59-29a0l et seq. (1994); Minnesota *(Minn. Stat. Ann.*, ch. 253B (1994 & Supp. 1996-1997); New Jersey (1994, L. 1994, c. 134, sec. 1); (North Dakota (1977, House Bill 1047); Wisconsin (*Wis. Stat.* sec 980.01 et seq. (1996). See also: Roxanne Lieb, Washington's Sexually Violent Predator Law: Legislative History and Comparisons with Other States. *Washington State Institute For Public Policy* (December 1996).
14. *Wash. Rev. Code* sec. 71. 09. 010 (West 1975 & Supp. 1991).
15. See *Wash. Rev. Code.* sec 71.09.030 (Supp. 1990-91) and *Kansas Statutes Annotated* 59-29A (1994).
16. American Psychiatric Association: *Diagnostic and Statistical Manual of Mental Disorders.* Fourth Edition. Washington, DC, American Psychiatric Association, 1994.
17. See Morse, id.
18. See note #15.
19. See Robert M. Wettstein: A Psychiatric Perspective On Washington's Sexually Violent Predators Statute. 15 *University Of Puget Sound Law Review* 587, 601-608 (Spring 1992).
20. See Stephen J. Morse, Culpability and Control, 142 U. Pa. L. Rev. 1587 (1994) for a discussion on the role of control in culpability for one's acts.
21. *Kansas v. Hendricks*, 117 S. Ct. 1997.
22. See In Re Young, 857 P. 2d 989 (Wash. 1993) appealed to federal district court in, *Young v. Weston*, 898 F. Supp. 744 (W.D. Wash. 1995); *State v. Post* and *State v. Oldakowski*, 197 Wis. 2d 279 (December 1995); In Re Blodgett, 510 N.W. 910 (1994), see federal court challenge in *Nicolaison v. Erickson*, 65 F. 3d 109 (1995) upheld Minnesota law; *People v. Superior Court* (Cain)

(1996) 49 Cal. App. 4th., see also *People v. Putney*, 1997. CA 726; In the matter of D.C., 1996 N.J. 351.

23. *In re Care and Treatment of Hendricks*, 912 P.2d 129 (Kan. 1996).
24. *Foucha v. Louisiana,* 504 U.S. 71 (1992). Case citing *Addington v. Texas,*_441 U.S. 418 (1979) which held among other things that an individual must suffer from something more serious than "idiosyncratic behavior" to justify loss of liberty. Also cited, was *O'Connor v. Donaldson,* 422 U.S. 563 (1975) in which the U.S. Supreme Court first held that a person who is mentally ill has a "right to liberty", and that right may not be abridged indefinitely on the sole basis of his mental illness. A state must prove in addition that an individual is likely to harm himself or others in addition to being "mentally ill".
25. See Meaghan Downey Kilebuck: Commentary: Kansas v. Hendricks: Is It Time To Lock The Door And Throw Away The Key For Sexually Violent Predators? 14 *J Contemp H L & Pol'y* 537 (1998).
26. *Kansas v. Hendricks*, 117 S. Ct. 2072 (1997).
27. See *Kansas v. Hendricks,* 116 S. Ct. 2072 (1997).
28. *APA Task Force Report On Sexually Dangerous Offenders*, American Psychiatric Association, publication scheduled for Spring, 1999. See pages 24-26.
29. See footnote # 22.
30. See Bruce J. Winick: Special Theme: Sex Offenders: Scientific, Legal, and Policy Perspective: Sexually Violent Predator Laws and Registration and Community Notification Laws: Policy Analysis: Sex Offender Law in the 1990s: A Therapeutic Jurisprudence Analysis. 4 *Psych. Pub. Pol. & L.* 505 (March/June, 1998).
31. Kansas v. Crane (No. 00-957. Argued October 30, 2001-Decided January 22, 2002) 269 Kan. 578, 7 P. 3d. 285, vacated and remanded (2002).
32. *United States v. Hinckley,* Crim. No. 81-306 (D.D.C. June 21, 1982).
33. Ralph Slovenko, *Psychiatry and Criminal Culpability* 25 (1995) also see: Stephen Morse, Culpability and Control, 142 U. PA. L. REV. 1587 (1994).
34. *Am. Psychiatric Ass'n, Statement On The Insanity Defense* 12 (1982).
35. See John Monahan, *The Clinical Prediction Of Violent Behavior* (1981); also, Christopher D. Webster & Robert J. Menzies, The Clinical Prediction of Dangerousness in *Law & Mental Health: International Perspectives*, 158-208 (David N. Weisstub ed., 1987).
36. Note that often epidemiological information is directly introduced into evidence on issues like dangerousness. However, at the Supreme Court level, epidemiologic data has been introduced to question such issues as the reliability of psychiatric diagnosis and the ability of mental health professionals to predict dangerousness. *O'Connor v. Donaldson,* 422 U.S. 585 (1974) relied on studies that showed the tentativeness of psychiatric diagnosis; *Barefoot v. Estelle*, 103 S.Ct.3383 (1983). (see esp. 3396-3399) which addressed the ability of psychiatrists to predict dangerousness; and *Addington v. Texas,* 99 S.Ct. 1804 (1979) (see esp. 1811) explicitly cited the "lack of certainty and fallibility of psychiatric diagnosis and prediction of dangerousness".
37. See Bruce G. Link & Ann Stueve, Psychotic Symptoms and the Violent/Illegal Behavior of Mental Patients Compared to Community Controls, in *Violence And Mental Disorders: Developments In Risk Assessment* (John Monahan and Henry J. Steadmen eds., 1994).
38. See Charles W. Lidz et. al.: The Accuracy of Prediction of Violence to Others, 269 *JAMA* 1007 1993).
39. Douglas Mossman: Assessing Predictions of Violence: Being Accurate about Accuracy. *62 J. Consulting & Clinical Psychol.* 783 (1994).
40. See Robyn M. Dawes et. al.: Clinical Versus Actuarial Judgment, 243 *Science* 1668 (1989).
41. See David B. Villeneuve & Vernon L. Quinsey: Predictors of General and Violent Recidivism Among Mentally Disordered Inmates, 22 *Crim. Just. & Behav.* 397 (1995).
42. Robert Menzies & Christopher D. Webster: Construction and Validation of Risk Assessments in a Six-Year Follow-Up of Forensic Patients: A Tridimensional Analysis, 63 *J. Consulting & Clinical Psychol.* 766 (see pages 775-777) (19d95).
43. John Q. La Fond: Special Theme: Sex Offenders: Scientific, Legal, and Policy Perspective: Forward: Sex Offenders and the Law. 4 *Psych. Pub. Pol. & L.* 3 :(March/June, 1998).
44. See R. Karl Hanson: What Do We Know About Sex Offender Risk Assessment? 4 *Psych., Pub. Pol'y, & Law* 50-72 (1998).
45. Gordon C.N. Hall: Sexual Offender Recidivism Revisited: A Meta-Analysis of Recent Treatment Studies, 63 J. *Consulting & Clinical Psychol.* 802 (1995).

46. Steven D. Gottfredson & Don M. Gottfredson: Behavioral Prediction and the Problem of Incapacitation 32 *Criminology* 441 see page 468 (1994).
47. Judith V. Becker & William D. Murphy: What We Know and Do Not Know About Assessing and Treating Sex Offenders, 4 Psych. Pub. Pol'y, & Law 116-137 (1998).
48. See V.L. Quinsey, M.E. Rice & G.T. Harris: Actuarial Prediction of Sexual Recidivism. 10 *Journal of Interpersonal Violence* 85-105 (1995).
49. Grant T. Harris & Marnie E. Rice: Special Theme: Sex Offenders: Scientific, Legal and Policy Perspective: The Science of Sex Offenders: Risk Assessment, Treatment, and Prevention: Appraisal and Management of Risk in Sexual Aggressors: Implications for Criminal Justice Policy. 4 *Psych. Pub. Pol. & L.* 73 (March/June, 1998).
50. Id. 8.
51. M.E. Rice, G.T. Harris & V.L. Quinsey: A Follow-up of Rapists Assessed in a Maximum Security Psychiatric Facility 5 *Journal Of Interpersonal Violence* 5, 435-448 (1990).
52. R.K. Hanson & M.T. Bussiere: Predicting Relapse: a Meta-analysis of Sexual Offender Recidivism studies. 63 *Journal Of Consulting And Clinical Psychology* 66, 348-362 (1998).
53. See Grant T. Harris and Marnie E. Rice, endnote #51.
54. See E.S. Janus & P.E. Meehl: Assessing the Legal Standard for Prediction of Dangerousness in Sex Offender Commitment. 4 *Psych. Pub. Pol. & L.* 297-322 (1998). Authors point out that report base rates for prediction are often low which precludes an efficient prediction instruments even within the limitations of currently employed instruments.
55. Grant T. Harris & Marnie E. Rice, footnote 53.
56. See J.J. Romero & L.M. Williams Group Psychotherapy and Intensive Probation Supervision with Sex Offenders. 47 *Federal Probation* 36-42 (1983).
57. See Harris & Rice, endnote #51, p. 13.
58. See Harris & Rice, endnote #51, p. 13.
59. See Harris & Rice, endnote #51, p. 14. Also J. P. Fedoroff, R. Wisner-Carlson, S. Dean, & F.S. Berlin: Medroxy-progesterone Acetate in the Treatment of Paraphilic Sexual Disorders. 18 *Journal Of Offender Rehabilitation* 109-123 (1992).
60. See Bruce J. Winick: Special Theme: Sex Offenders: Scientific, Legal, and Policy Perspective: Sexually Violent Predator Laws and Registration and Community Notification Laws: Policy Analysis: Sex Offender law in the 1990s: A therapeutic Jurisprudence Analysis. 4 *Psych. Pub. Pol. & L.* 505 (March/June, 1998) for an excellent discussion on therapeutic jurisprudence. See also David B. Wexler & Bruce J. Winick: *Essays in Therapeutic Jurisprudence*, 1991.

CHAPTER 6

CRIMINAL BEHAVIOR AND BRAIN IMAGING TECHNO-SCIENCE

Midnight, March 13th, 1995. Jerry Kurtin, a 6' 3" 240 lb former high school football star left his video game to buy cigarettes at a Seven-Eleven store at the corner of Kirby and Holcombe Boulevard in Houston, Texas. As he approached the counter he saw a thin disheveled elderly man negotiating the price of a pastry with the clerk. The customer's voice, becoming ever louder, delivered a flurry of obscenities at the clerk. Jerry told him to calm down, and that nothing would come of his use of obscenities. After a moment of silence, the man turned to Jerry and called him "a big lug".

Jerry insisted that the man settle down, which appeared to have some effect; but then the man pointed his right index finger at the clerk coming within inches of her nose. Jerry grabbed him around his upper arms and pulled him away from the counter. Released, the man attempted a swing at Jerry and then began throwing cans of vegetables frantically gathered from a nearby shelf. This effort came to naught, but he did succeed in rousing Jerry to a level of violence they would both regret. Jerry recalled that at that moment "something clicked" in him and he ran full speed at his tormentor, baring his shoulder much as would a lineman on a football team, pushing his adversary forcefully against the glass refrigerator doors that lined one of the walls in the store. Blood oozing from his mouth, the man collapsed to the floor. Jerry, unable to calm the screaming clerk, impulsively ran after her, hitting her with his body and rendering her unconscious.

The police, having been notified earlier by the clerk who had pressed a button under the counter, arrived as Jerry was leaving the store. They arrested Jerry, and called for an ambulance, which took both of the injured persons to the Herman Hospital emergency room. At time of arrival, however, the elderly man was dead and the woman still unconscious having suffered a concussion.

BRAIN IMAGING TECHNO-SCIENCE AND HIERARCHIES OF INFERENCE

At trial the defense attorney contended that Jerry suffered from a biological brain disorder which prevented him from reasonably controlling his emotions and behavioral responses to provocation. A neuroscientist hired by the defense was called to the stand and initially discussed "neuro-imaging", a relatively new technique which involves several technologies (1). These approaches include Computerized Tomography (CT scan) and Magnetic Resonance Imaging (MRI) which can reveal structural or morphological changes in the brain. The CT Scan produces a picture based on the degree to which different densities of brain tissue absorb and deflect X-ray beams, whereas MRI images are constructed from electromagnetic signals. The subject is placed in the strong external magnetic field of the scanner, which creates signals emitted by the proton nuclei of hydrogen atoms (present in tissue water) that are pulsed with radio frequency waves. These signals differ depending on the nature of the tissue examined and are fed into a computer which creates a magnetic resonance image of the brain (2). In contrast, PET (positron emission tomography) scans and SPECT (single emission tomography) scans, can detect abnormalities or differences in brain blood flow or metabolism, whereas CEEGs (computerized electroencephalograms) can identify abnormalities in electrical potentials in the brain. PET involves positron-emitting radioisotopes which label glucose, water, and other chemical agents including medications (3). A radiolabeled compound is injected into a subject's vein. Several minutes later the PET scanner detects the spatial distribution of these isotopes in the brain and feeds the radiation counts in the various brain locations through a computer which determines the metabolic rate, and/or difference in blood flow of specific brain regions, translating these into a simulated image of the brain. The regional differences in metabolic activity are projected in color variations from yellow and red (high activity) to blue and green (low activity) in computer generated cross-sectional images of the brain. So it is possible to give a subject (while under the scanner) a task, such as listening to a Bach cantata, and determine which region(s) of the brain is activated by the subject's response. SPECT is also capable of providing information on brain function, but the spatial resolution is poor in comparison to PET. Finally, functional MRI (fMRI) has been added to the list of imaging technologies that can detect brain activity (4). This technology involves superimposing a time series of changes in the oxygenation of blood, which is reflective of metabolic activity, upon a structural MRI image of the brain. Though relatively new, this technology is being refined and is increasingly used for studies of cognition and behavior. It has the advantage over PET of not injecting a radioactive substance into the subject. Nonetheless, PET scans provide better resolution and remain the current "gold standard" for assessing brain function (5).

After describing the technologies, the neuroscientist reviewed the results of PET studies involving radiolabeled glucose in Jerry Kurtin. Projecting the slides of Kurtin's scans next to those of a composite age-and sex-matched control of similar handedness (Kurtin was right handed) showed differences which this expert claimed showed a biological abnormality in Kurtin's brain (6). At baseline--in the resting state with no stimulation--the control revealed symmetry between the right and left sides of the brain. The color components were mostly blue-green, with areas of activity in red and yellow, particularly at the temporal lobes and basal ganglia, amygdala and cingulate gyrus areas of the brain. Kurtin's scan at baseline showed comparable functioning on the right side of the brain but with hypometabolism on the left, with loss of color coordination between the two sides and differences that, though not marked, supported the relative lack of activity on the left side. In addition, as compared to the control study, Kurtin showed generalized lower prefrontal glucose metabolism. Research on the autonomic nervous system--through studies of skin conduction and heart rate--of antisocial and aggressive behavior in children and adolescents has shown that under-arousal is closely linked to the uninhibited temperament found in juvenile delinquency and adult aggressive behavior. In addition, paradoxically, the link between autonomic nervous system functioning and violence seems strongest in those offenders who come from relatively benign home backgrounds (7). Subsequent studies have shown that murderers without a past history of psychosocial deprivation, such as would be seen in childhood neglect and abuse, show lower prefrontal glucose metabolism than in controls or even in murderers who have had early psychosocial deprivation (8). It has been argued that this class of murderers with prefrontal deficits possess little if any social drive to violence but rather have relatively powerful subcortically mediated predispositions to violent behavior.

The neuroscientist followed the baseline studies with scans taken during stimulation-provocation with a loud sound (9), demonstrating that Kurtin's scans were markedly different from the control. Kurtin's images showed striking hypometabolism of the left temporal lobe region during stimulation which, according to the neuroscientist indicated that Kurtin lacked the ability to control his aggressive responses to provocation. Furthermore, the response was said to provide the basis for Kurtin's potential for recidivistic violence, described earlier in the trial by prosecution witnesses who detailed episodes of aggressive violence by Kurtin since his teenage years. None of these incidents was serious or a cause of permanent damage to the victims. Kurtin's status as a successful high school football player appeared to figure prominently in protecting him against legal action.

The neuroscientist then argued that Kurtin had a biological proclivity to violent responses as well as the inability to control such responses once triggered, and he quoted several earlier studies with PET. One involved studying the glucose metabolism of murderers who pleaded not guilty by reason of insanity (NGRI) compared with 41 age- and sex-matched controls (10). This study, using a continuous performance challenge task, demonstrated that the murderers had abnormal asymmetries of activity of the left hemisphere over the right. Reduced glucose metabolism was present in various brain structures on the left side which have been shown in animal studies to be associated with violent behavior, indicating a network of abnormal brain processes in the cortical and subcortical regions that predispose to violence (11). These areas included the prefrontal cortex, superior parietal gyrus, amygdala, thalamus and medial temporal lobe.

The neuroscientist brought by the Prosecution then took the stand and drew attention to the conceptual problems with imaging technologies and indicated reasons why Kurtin's scans should not be admitted for the purposes of excusing his behavior. The judge, based on his empowerment as gatekeeper through the Daubert decision (12), disallowed the introduction of Kurtin's PET studies to the jury in their consideration of the appropriateness of extenuating and exonerating circumstances for his murderous act. As a result Kurtin was found guilty of 2nd degree murder and felonious assault. When the sentencing phase occurred, the defense once again introduced information from the brain scans to argue against the death penalty, which was proposed by the Prosecution. The defense claim was that the biological condition of Kurtin's brain impacted most powerfully on his inability to control murderous impulses, and that therefore he did not meet the standard of malicious intent or wanton disregard of human life to warrant the death penalty. At this phase the judge allowed the PET Scan results to be introduced on behalf of Mr. Kurtin, who was given life imprisonment instead of the death sentence.

IMAGING TECHNOLOGIES IN THE ARENA OF THE COURT

Although the Kurtin case was tried under criminal law, brain scans have been introduced in civil cases (13), such as tort cases involving construction, traffic or other kinds of accidents (14). Brain scans have been introduced in other criminal cases as well (15). Most noteworthy of these was the Hinckley case in which structural changes were demonstrated to explain violent behavior. During this 1981 trial, the District of Columbia court allowed the defense to put a psychiatrist on the stand (16). Using the image of Hinckley's brain produced by a CT scan, the psychiatrist pointed to widening of the sulci in his brain, claiming that this substantiated a diagnosis of schizophrenia. However, there is no scientific justification for claiming that this finding bears has any relation to schizophrenia, and, in fact, widened sulci are commonly found in normal individuals. Furthermore, the brain scan did not show enlarged ventricles which have been shown to correlate with schizophrenia.

Brain scans for demonstrating structure, such as CT Scans and MRIs, have been used successfully to support a defense. In the diminished capacity defense of United Way executive William Aramony in 1995, an MRI scan played a pivotal role on his behalf (17). Aramony had embezzled hundreds of thousands of dollars from the charitable foundation for his lavish lifestyle. His attorney claimed that Aramony was incapable of developing the requisite criminal intent for embezzlement. The reason: his brain had been "shrinking" during most of the late 80's and 90's, which could be substantiated by an MRI scan that the defense intended to introduce. If accepted by the jury, the diminished capacity defense could have brought about an acquittal. Fearing the persuasive effects on the jury of having a neuro-image presented in court--even though the "shrinking brain" thesis had little scientific basis--the prosecution agreed to a favorable plea bargain for the defense.

Functional brain scans, such as PET, SPECT and fMRI, have also been admitted into evidence in both criminal and civil cases. In criminal law, PET scans have been admitted since the 1980's with images from PET scans in several criminal cases (18). The overall impression from this experience was that neuro-images provide very persuasive evidence (as it is perceived by juries). In 1992, a case in New York explicitly addressed the admissibility of PET scans in the court (19). The accused, a retired advertising executive, strangled his wife, during an argument. He then threw her body from their 12th floor apartment in Manhattan, believing this would give the appearance of a suicidal act.

At trial the defense introduced a psychiatrist who testified that at the moment of the murder the defendant lacked the cognitive ability to understand the nature and consequence of his act as well as the fact that it was wrong (20). He was prepared to add that the cognitive power of the accused was compromised by organic brain damage. A PET scan confirmed an earlier MRI finding of a cyst in the arachnoid membrane. Furthermore, the PET scan pointed to metabolic imbalances in areas near the cyst and opposite it (21). The first issue before the court was the appropriateness of allowing PET scans into evidence. A PET expert supported the defense position claiming that PET is generally accepted as a technology for measuring glucose metabolism in the brain. Another expert acknowledged that the defendant's brain was abnormal due to the cyst which compressed the brain and therefore made it appear abnormal (22). Apparently the regions of the brain which appeared most abnormal in the PET scan were the frontal lobes, which are known to be involved with executive functions, i.e. ability to reason, to plan, judgment, insight. However, the expert disagreed with the contention that the data were sufficient to link these metabolic abnormalities on the PET scan with violent behavior.

The judge noted one of the studies cited (Volkow and Tancredi), which reported on four violent patients all of whom showed hypometabolism of the frontal lobes on PET (23). The expert emphasized that these findings were potentially of interest, but by no means conclusive of this relationship (24). As a result, based on the Frye test (25) which controlled the admissibility of evidence in the New York courts, the PET scans, including the presence of the arachnoid cyst and metabolic imbalances, were admitted into evidence to support the psychiatric diagnosis; they were not admitted to establish evidence of a mental disease or defect, or to establish a relationship between the scan results (particularly the hypometabolism of the frontal lobes and the arachnoid cyst) and violent behavior (26). On the other hand, the impact of the court's decision was significant. By allowing the PET scan to be introduced in support of the insanity defense, the prosecution allegedly was motivated to accept a manslaughter plea (27).

Even though the court restricted the nature of the admissibility of neuro-imaging for the above case, the issue nonetheless caused considerable controversy. One expert expressed alarm by what she viewed as a legal precedent insisting that a PET scan cannot predict the behavior of a person through nonspecific images (28); she also pointed out that abnormalities of the frontal lobes can be seen in a variety of conditions--strokes, head trauma, schizophrenias, mood disorders--without concomitant violence (29).

In several subsequent criminal cases in New York state we find a different position on the admissibility of PET information. Both of these cases involved crimes that occurred in New York State. In People v. Jones (30), a woman, screaming and cursing while breaking into the defendant's home by using a large piece of lumber to demolish the bedroom window was shot and killed by the defendant. The defendant was aware that the victim was physically larger than he. At trial, the defendant was precluded from introducing neurological tests or brain scans to support his defense justification; he argued that as a child he had sustained a traumatic head injury that caused permanent brain damage, which was compounded by a 30 year history of alcoholism. Both of these conditions caused cognitive limitations affecting his ability to think quickly and flexibly. The expert's testimony, had it included brain scans, would have addressed more than the defendant's ability to form intent; it would also had revealed Jones' ability to perceive risk, crucial to the defense for justification of the shooting (or so the defense would have argued).

The appellate court reversed the homicide conviction because of the state's failure to admit brain scans.

Another case, decided in the Federal court of the Eastern District of New York, involved Vincent Gigante, leader of a Mafia crime family (31). Brought to trial for numerous conspiracy and RICO (Racketeer Influenced and Corrupt Organization Act) charges and found guilty by a jury, he alleged that he was incompetent to be tried and sentenced and therefore should be institutionalized and treated in an appropriate facility. Experts for the defense, all well known authorities, included one who had done a PET scan on the defendant and who had concluded that Gigante had an abnormal metabolic pattern on the scan typical of patients with dementia. He further claimed that the pattern was not the product of psychopharmacological drug action (32). Other experts, including a neuropsychologist, stated that these findings were confirmed by their own battery of tests. The prosecution had three witnesses, one of whom argued against the validity of the PET Scan results and the neuropsychological and clinical assessments of Alzheimer's disease or multi-infarct dementia, claiming that these scans (two earlier SPECT scans, a CT scan, and a PET scan) revealed abnormalities that were not consistent with the clinical impression of vascular dementia; he also claimed that Mr. Gigante's prescribed psychotropic medications would have a profound impact on the PET images, adding that the metabolic presentation by PET scans was not consistent with a dementia severe enough to correlate with clinical signs. The prosecution, then, took the position that the PET scan results were inaccurate because of his medication which could distort the results, and because the control group was not receiving the same medication(s) and they differed significantly in age with regard to the defendant.

On the basis of all the testimony the judge concluded that the defendant's experts and test results were unreliable and that therefore PET scans, presumed to demonstrate that Gigante suffered from organic brain disorder were not to be admitted as evidence at the sentencing.

The PET scan results in the Kurtin case were not admissible at either the competency hearing or at trial for the insanity defense. However, at the sentencing hearing or the penalty phase of the trial, as we indicated earlier, the PET images were admitted into evidence. Usually, testimony involving neuro-images is largely confined, in most states, to sentencing proceedings. Many of these cases, particularly in California, Florida and Texas involve cases in which the death penalty is recommended. In a California case (People v. McNamara) (33) which involved a mass murderer, a PET scan was offered as evidence to claim that McNamara suffered from an organic brain condition. After hearing this evidence the jury rejected the death penalty. Subsequently, the defense attorneys claimed that the jurors acknowledged that they gave significant credence to the brain images.

Two Florida cases (34,35) further illustrate how these scans can become part of the defense strategy in influencing a jury against the death penalty. In the first case, Davis v. State, the defendant was denied a PET Scan, which he requested for post-conviction relief of his death penalty sentence. The defendant claimed that the PET scan did not exist when his triple homicide trial occurred in 1983 and requested that a PET scan be ordered to determine if he had a seizure disorder which affected his competency at the time of the murders. The Florida Supreme Court denied this request, stating that the defendant's claim was highly speculative, particularly since the defendant has previously informed his attorney that he remembered the killings, which would not be the case if he suffered a seizure.

However, in Hoskins v. State (35) the defendant moved that he should be transported to another county in Florida to obtain a PET scan since the expert for the defense claimed that the test was necessary for a proper assessment of the defendant's frontal lobe ability and impulse control. The Florida Supreme Court ruled that due process required this test be conducted, for when a criminal defendant can demonstrate that his mental status is at issue, he must be given access to appropriate tests. The testing was done and as a result the Supreme Court vacated the death sentence.

Neuro-imaging technology, particularly PET Scans, are being used increasingly to study receptor sites in the brain and the effects of drugs on brain function. In the Florida case of Ted Harris (36) the results of PET research combined with other biological studies of the brain were very important in overturning the death penalty. During a murder this defendant, a cocaine abuser, had stabbed an elderly lady over sixty times. Though on cocaine at the time, he was convicted of homicide and given the death penalty. Subsequently a New York city law firm took the case pro bono to argue against the death penalty, using an argument based on dopamine dysregulation of cerebral areas involved in channeling of drive and affect leading to loss of control and compulsive behavior (37). The cingulate, orbital and prefrontal cortices of the brain are associated with hypometabolism in cocaine abusers (36,37). These structures of the brain are involved in regulation of "drive" (38) and their dysfunction could lead to the inability of an abuser to restrain from using cocaine--i.e. compulsive drug administration. The literature indicates that destruction of the orbito-frontal cortex or mesocortical dopamine pathway can lead to the emergence of repetitive behavior that cannot be terminated (39). On the other hand, in patients with classical obsessive-compulsive disorders, greater metabolic activity of the orbito-frontal cortex and basal ganglia has also been reported (40). Thus, the orbito-frontal cortex and basal ganglia form part of the circuit that presumably plays a critical role in the occurrence of repetitive behaviors in patients with obsessive-compulsive disorder (41). Animal studies have shown that lesions of this circuit lead to the emergence of repetitive behaviors that cannot be terminated by external or internal cues (42).

Studies introduced at the Harris trial indicated that under certain conditions cocaine increases the compulsive reactions and aggressive behavior of animals (43). In clinical practice it has been observed that violent behaviors increase with cocaine use (44). The types of violent behavior reported range from minor psychological aggressions to major physical acts that include murder and rape. Pharmacological mechanisms underlying the induction of cocaine-associated violence may involve dopamine and serotonin neurotransmitter systems affected by cocaine. The court was informed that cocaine acts in those areas of the brain, especially the limbic system, that subserve aggressive and violent behaviors (45). Augmenting the claim that cocaine creates dopamine dysregulation, leading to loss of control and compulsive behavior, was evidence showing the close linkage between cocaine and problems with judgment, paranoia and psychosis. The effect on judgment is fueled by cocaine-induced suspiciousness, persecutory fears, paranoia, anxiety and hypervigilance. Cocaine users may be excited, disinhibited and hyperactive (46). They may become suspicious and paranoid within a few hours of initiating the use of high doses of cocaine (47). Chronic use of a stimulant like cocaine can produce a paranoid psychosis that is similar to acute paranoid schizophrenia. A study (48) involving the evaluation of 55 individuals admitted to a hospital for treatment of cocaine dependence showed that at least 53% experienced transient cocaine-induced psychosis and that males were significantly more likely than females to develop psychosis. There was no significant difference in lifetime amount of cocaine use, or the amount of cocaine use in the month before admission between those who experienced psychosis and those who did not. However, the psychosis-positive group used significantly more cocaine in the year prior to admission. Another study, of cocaine abusers and paranoid schizophrenics, showed that the latter had more bizarre delusions but that both groups experienced command hallucinations (49). Also, violent behavior was common among cocaine patients, especially those with psychosis (50).

As seen in the Harris case, a survey of biological studies using PET is far more easily admitted into evidence than a PET image of a specific defendant. There are several reasons for this. First, the PET studies on brain biology are subjected to rigorous statistical analysis and published in respected neuroscience journals. As a result they have been evaluated and approved for publication by other scientists (peer reviewed) giving the data an authoritative note in keeping with the Frye and Daubert tests. Secondly, the PET image of a particular defendant has a powerful, potentially biasing effect in the court room. The image creates a reality for the jurors, frequently well beyond its true scientific merit. The concern has been that the "image", even if not predicated on powerful statistical or scientific grounds, empowers the expert's testimony in a way a clinical assessment through interview cannot. Judges tend to carefully weigh admittance of such images in court trials. Thus, paradoxically, PET studies of the brain and behavior packaged in academic papers are generally admissible, whereas a PET study without the status of publication in a peer-reviewed journal is subject to claims, for example, that the science is not "refined", or that what the study purports to demonstrate is based on interpretation that may be prejudicial, and that the image of parts of the brain are basically nonspecific.

READING MEANINGS INTO IMAGES

The image created by a PET, MRI, SPECT, or CEEG is used as a "fact" in the courtroom. This fact must stand up to various tests to justify its role in determining the validity of a position in a legal dispute, and the problematics of PET scans as "facts" have been discussed. Some experts insist that PET images are too nonspecific and cannot predict behaviors. For example, as noted previously, frontal lobe hypometabolism can accompany a variety of conditions such as stroke, head trauma, mood disorder or violence proneness.

Several considerations should precede reflections on what neuro-images represent in the brain and how they relate to thought, mood and behavior. First, to be useful, the image must have sufficient sensitivity to detect the biological condition of the brain (51). During a PET scan a defendant is given a positron-emitting compound which the camera records and feeds to the computer that creates the image. The sensitivity of the instrument to detect what is occurring in specific regions of the brain is essential. For example, if it is claimed that a PET image of the amygdala reveals hypometabolism in people biologically disposed to violence, then a difference should be detected (in a violent offender) that is significant compared to matched normal studies (at least one standard deviation outside normal limits).

Next, one must address the specificity of the tests and images. The abnormality revealed by PET and simulated in an image must be consistent with actual behavior, e.g., a PET expert is shown an image of the brain of a pedophile. The basal ganglia and cingulate gyrus are hypometabolic, suggesting that the defendant suffers from a biological condition. Does the image actually represent a biological condition? Is this the only possible explanation, or could it be merely an artifact of the test itself under certain circumstances? Would the image be different if taken at another time?

Some question the specificity of PET images, claiming that the images, e.g. of the hypometabolism of the frontal lobes, are nonspecific. A second related problem is when an abnormal activation pattern is detected and imputed to be the cause of a specific biological abnormality and concomitant behavior. That linkage of causation may be artifactual. For example a violent person may show three or four different activation patterns in the brain. One may appear all the time during studies, but be unrelated to the cause of violence.

Additionally, one must address the predictive value of the neuro-image: does a particular activation pattern indicate that a specific behavior or behavioral pattern is likely to occur? If so, this would allow us to assume a direct relationship between the image and brain tissue pathology, and ultimately a specific deficit and/or behavioral abnormality. At the same time it would allow us to assume that other conditions or behaviors are most unlikely with a specific image. Reliability of a test of structure and function is very important and particularly difficult with psychological or behavioral conditions. For over half a century it has been known that inconsistency exists between brain structure during autopsies and behavior; individuals with severe behavioral abnormalities exhibited only minimal structural differences. This structural test would, therefore, create a range of "false" negatives, i.e., failure to detect a condition (the behavioral abnormality). More recent studies of EEGs which revealed massive brain alterations in patients with minimal behavior disorders produced similar results (51). But in this case, the EEG pattern is detecting false positives.

Examination of sensitivity, specificity, predictability, validity, and reliability of neuro-imaging technologies reveals the inaccuracies in these structural and functional tests of brain function and behavior. However, PET and other studies reported in peer-reviewed journals in which statistical methods are used take on greater significance than individual studies for the reasons described above. But, paradoxically, if the claims of lack of sensitivity and specificity are valid, then the mere fact that PET studies are framed in a statistical system does not validate their explanatory power for brain structure and behavior. Nevertheless, for the social system--particularly the courts--such large-scale studies take on greater importance because of their legitimacy in a peer reviewed journal. Also, claims that are made about neuro-imaging technologies, i.e. regarding their accuracy as producers of "fact", can also be made about bioenzyme studies, or virtually any diagnostic test. Psychoanalytical interpretations have until relatively recently been very powerful in criminal cases. Yet the assumptions behind these interpretations have been virtually untestable. Hence, what emerges from this exercise is this: there is a basic difficulty in establishing a clear definition of a scientific fact and also in cataloging that fact in a world of similarly imprecise facts so that the information can be used in making decisions in a court of law where they have important social consequences.

NOTES

1. Brodie, J. "Imaging for the clinical psychiatrist: facts, fantasies and other musings. *American Journal of Psychiatry* 153:145 (February 1996).
2. Andreasen, N.C. (ed) *Brain Imaging: Applications in Psychiatry,* 1989. See in particular J.A. Coffman, "Computed Tomography in Psychiatry", p. 17-36; and N.C. Andreasen, "Nuclear Magnetic Resonance Imaging", p.67 ff.
3. Volkow, N.; Tancredi, L: Positron emission tomography: a technology assessment. *International Journal of Technology Assessment in Health Care* 2:577-594 (1986).
4. See Brodie, supra note 1.
5. Morgan, A.E., Brodie, J.D., Dewey, S.L.: What are we measuring with PET? *Quarterly Journal of Nuclear Medicine* 42:151 (Sep. 1998).
6. See Volkow, N., Tancredi L.R.: Neural substrates of violent behavior: a preliminary study with positron emission tomography. *British Journal of Psychiatry* 151:668-673 (1987).
7. See Raine A., Phil D, Stoddard J, Bihrle S, Bucksbaum M: Prefrontal glucose deficits in murderers lacking psychosocial deprivation. *Neuropsychiatry, Neuropsychology and Behavioral Neurology* 11:1 (1998).
8. See Raine et. Al. Note 7 above.
9. See Tancredi, L.R., Volkow, N.: A theory of the mind/brain dichotomy with special reference to the contribution of positron emission tomography. *Perspectives in Biology and Medicine* 35:549 (Summer 1992).
10. Raine A., Buchsbaum M., LaCasse L: Brain abnormalities in murderers indicated by positron emission tomography. *Biological Psychiatry* 42:495 (Sep 1997).
11. See Volkow N, Tancredi L, supra note 6.
12. *Daubert v. Merrell Dow Pharmaceutical, Inc, 509 U.S. 579 (1993).*
13. Brain scans including PET are becoming more frequent issues in court cases at both the Federal and state levels. The following are some of the Federal cases where the admissibility of a neuroimage became important. The numbers of cases at a state level are numerous:
    Lockett v. Anderson, 230 F.3d 695 (5th Cir. 2000)
    Jackson v. Calderon, 211 F.3d.1148 (9th Cir. 1999), rehearing denied Dec. 8, 1999; certiorari denied January 8, 2001, 2001 U.S. Lexis 161.
    United States v. Gigante, 166 F.3d. 75 (2nd Cir. 1999) Certiorari denied January 18, 2000, 200 U.S. Lexis 563.
14. *Hose v. Chicago Northwestern Transp. Co., 70 F.3d 968 (8th Cir. 1995).* See D'Agincourt, L: PET findings support insanity defense case. *Diagnostic Imaging.* January 1993:45 In interview with Dr. Abass Alavi, he noted that CT is the primary technology to assess head trauma, but that he has used PET in cases, such as one involving a woman claiming memory loss from a car accident, because it shows metabolic disturbances suggestive of axonal injury, and therefore provides something more than structural abnormalities.
15. The trend in recent years towards using neuroimaging on psychiatric conditions in the courtroom appears to be increasing. Concern about the use of these technologies has even resulted in medical professional organizations making recommendations to the court. See: Society of Nuclear Medicine Brain Imaging Council: Ethical clinical practice of functional brain imaging, *Journal of Nuclear Medicine* 37:1256 (1996). The Society recommends that neuroimaging reports for forensic purposes include specific assessment of the quality of the scan, and the degree of compliance of the technique of scanning and methodology applied to a particular patient with the published research papers that identify and define the disease-specific patterns. This provides advice to the court as to the type of analysis it should conduct to assess technical adequacy of neuroimaging evidence.
16. Nelkin D; Tancredi L: *Dangerous Diagnostics: The Social Power of Biological Information.* University of Chicago, Chicago, 1994, 140-141.
17. See Kulynych J: Psychiatric neuroimaging evidence: A high-tech crystal ball? 49 *Stan. L. Rev.* 1249 (May 1997).
18. See Nelkin D; Tancredi L: supra note 16 pp. 156-157.
19. *People v. Weinstein, 591 N.Y.S. 715 (Misc. 1992).*
20. *People v. Weinstein* supra note #19, page 723.
21. *People v. Weinstein* supra note #19, page 718.
22. *People v. Weinstein* supra note #19, page722.

23. Volkow N; Tancredi L, supra note 6, page 668.
24. *People v. Weinstein* supra note #19, page 725. However subsequent studies by Raine corroborated these early preliminary studies. See: Raine A; Buchsbaum MS; Stanley J; Lottenberg S; Able L; Stoddard J: Selective reductions in prefrontal glucose metabolism in murderers, *Biological Psychiatry* 36:365 (1994) and Raine A; Meloy JR; Bihrle S; Stoddard J; LaCasse L; Buchsbaum MD: Reduced prefrontal and increased subcortical brain functioning assessed using positron emission tomography in predatory and affective murderers. *Behavioral Sciences & The Law* 16:309 (Summer 1998).
25. *Frye v. United States, 293 F. 1013 (D.C.Cir. 1923).*
26. *People v. Weinstein,* supra note #19, page 717 (Sections 13 and 16).
27. Sherman R: PET brain scan held admissible in murder case. *National Law Journal,* November 2, 1992:10.
28. Agincourt, L: supra note 13, 49.
29. Mayberg, H.S.: Functional brain scans as evidence in criminal court: an argument for caution. *Nuclear Medicine* 33:18N (1992). This perspective has also been supported by Dr. John Mazziotta who feels that research on the sensitivity, specificity, and overall utility of PET in forensic areas is lacking. See also, Mazziotta J.C.: Mapping human brain activity in vivo. *West Journal of Medicine* 161:273 (1994).
30. *People v. Jones, 620 N.Y.S. 2d 656 (1994).*
31. *United States v. Gigante, 982, F. Supp. 140 (E.D.N.Y. 1997).*
32. *United States v. Gigante,* supra note 31, see Monte S. Buchsbaum testimony.
33. *People v. McNamara, No. CV-88-5343-ER (C.D.Cal. 1990), aff'd, 923 F.2d 862 (9th Cir. 1990, certiorari denied, 501 U.S. 1220 (1991).*
34. *Davis v. State, 1999 Fla. Lexis 1133.*
35. *Hoskins v. State, 702 So. 2d 202 (Fla. 1997).*
36. Theodore Harris, *Appellant v. State of Florida,* 528 So. 2d. 361; 13 Fla L. Weekly 420 (1988).
37. Volkow N.D.; Fowler J.S.; Wang G-J; Hitzemann R.; Logan J.; Schlyer D.; Dewey S.; Wolf A.P.: Decreased dopamine D2 receptor availability is associated with reduced frontal metabolism in cocaine abusers. *Synapse* 14:169 (1993). Also more recently Volkow N.D.; Fowler J.S.: Addiction, a disease of compulsion and drive: involvement of the orbitofrontal cortex. *Cerebral Cortex* 10:318 (March 2000).
38. Stuss D.T.; Benson D.F.: *The Frontal Lobes.* Raven Press, N.Y. (1986) pp. 12-28.
39. Kolb B: Studies on the caudate putamen and the dorsomedial thalamic nucleus: implications for mammalian frontal lobe function. *Physiol. Behav.* 18:237 (1977); also, Le Moal M.; Simon H.: Mesocorticolimbic dopaminergic network: functional and regulatory notes. *Physiol. Rev.* 71:155 (1991).
40. Baxter L.R.; Phelps M.E.; Mazziotta J, et. Al.: Local cerebral glucose metabolic raters in obsessive-compulsive disorders. *Archives General Psychiatry* 44:211 (1987).
41. Modell J.G.; Mountz J.M.; Curtis G. et. al.: Neurophysiologic dysfunctions in basal obsessive compulsive disorder. *Journal Neuropsychiatry* 1:27 (1989).
42. Kolb B.: Studies on the caudate-putamen and the dorsomedial thalamic nucleus of the rat: implications for mammalian frontal lobe functions. *Physiol. Behav.* 18:237 (1977).
43. Darmani N.A.; Hadfield M.G.; Carter W.H. et. al.: Acute and chronic effects of cocaine on isolation-induced aggression in mice. *Psychopharmacology* 102:37 (1990).
44. Miller N.W.; Gold M.D.; Mahler J.C.: Violent behaviors associated with cocaine use: possible pharmacological mechanisms. *International Journal of Addictions* 26:1077 (1991); see also: Volkow N.D.; Fowler J.S.; Wang G-J: Imaging studies on the role of dopamine in cocaine reinforcement and addiction in humans. *Journal of Psychopharmacology* 13:337 (1999).
45. Goldstein P.J.; Belluci P.A.; Spunt B.J. et. al.: Frequency of cocaine use and violence: a comparison between men and women. *NIDA Research Monograph* 110:113 (1991).
46. Gilman A.G.; Rall T.W.; Nies A.S. et. al.: *Goodman and Gilman's, The Pharmacological Basis of Therapeutics* (8th ed.). Pergamon Press, 1990, 541-542.
47. Gilman A.G. et al, 544.
48. Brady K.T.; Lydiard R.B.; Malcom R. et. al.: Cocaine-Induced Psychosis. *Journal of Clinical Psychiatry* 52:509 (1991).
49. Mitchell J.; Vierkant A.D.: Delusions and hallucinations of cocaine abusers and paranoid schizophrenics: A comparative study. *Journal of Psychology* 125:301 (1991).

50. Manschreck T.C.; Laughery J.A.; Weisstein C.C. et. al.: Characteristics of freebase cocaine psychosis. *Yale Journal of Biology and Medicine* 61:115 (1988).
51. See Nelkin D.; Tancredi L. supra note 16.

CHAPTER 7

DNA FINGERPRINTING

There has been a growing public fascination with DNA based technologies, particularly for the identification of perpetrators and victims of crimes, as well as the use of DNA fingerprinting in exculpating a prisoner who has spent years in jail. Crimes committed before the era of DNA technology can also be reexamined. For example, a strand of hair taken from the back seat of a borrowed car driven the day of Hoffa's disappearance (1) on July 30, 1975 by his ostensible friend, Charles O'Brien, matched hair taken from Hoffa's hairbrush. By linking O'Brien to Hoffa the day he disappeared, this evidence may prove to be an important break in the case; at least it would exclude the suspect if there were no match.

Identification of criminals through DNA techniques is a relatively new development in criminal investigations. The first use of DNA in a crime occurred in Yarborough, England (2). In this case police arrested a local delivery boy, George Howard, for the rape and murder of two teenagers which had occurred in different years during the mid 1980s. Howard owned a red motorcycle and helmet that matched those items seen by witnesses at the site of the murders. During police interrogation Howard, who worked in an asylum for the criminally insane near where the murders occurred, confessed to one of the killings, stating that he "probably went mad and didn't know it (3)". Police were convinced that Howard's confession solved one of the murders, but to establish that he was responsible for the second murder, the police sent a sample of Howard's blood along with semen and blood samples collected at the crime scenes to Dr. Alec Jefferys who was studying DNA and who lived in the area. Using DNA "fingerprinting" Jefferys concluded that not only was Howard innocent of the first crime, but he was likewise not responsible for the one he confessed to committing. Impressed with the power and precision of identification, the police then used a "dragnet" approach (4). They required that blood samples be given by all the local men between 17 and 34 years of age in order to compare with the DNA obtained at the murder sites. Nearly 5,000 men were investigated and the DNA studies led to the capture and prosecution of serial rapist and murderer Colin Pitchfork, the first criminal identified and ultimately brought to justice based primarily on his DNA (5). Shortly thereafter in 1987, DNA evidence was introduced for the first time in the United States, in an Orlando Florida court. Tommy Lee Andrews was shown to have DNA that matched the DNA of the semen from a rape victim. Testimony stated that the probability that the DNA was not that of Andrews was one in ten billion. As a result, he was convicted and sentenced to 22 years in prison (6). Subsequently DNA evidence assumed an increasingly important role and it is estimated that, in this country, well over 10,000 criminal investigations each year involve DNA evidence. Interestingly, nearly one-third of all primary rape suspects are cleared of charges through DNA testing (7).

DNA testing has also been found to be useful in civil litigation and administrative law. It has been shown to be determinative in many family law disputes, such as paternity.

THE SCIENCE OF DNA EVIDENCE

Human cells contain amino acids, lipids, proteins, and carbohydrates, as well as nucleic acid (8). Chromosomes within the nucleus of the cell include proteins, and DNA (deoxyribonucleic acid) which carries the hereditary genes. Within the nucleus are twenty- three pairs of chromosomes (total 46), twenty-two of which are autosomal (non-sex) plus two sex chromosomes (female sex configuration XX and male XY). One half of the twenty-three chromosomal pairs are supplied by the mother and one half by the father. DNA (deoxyribonucleic acid) is a string (polymer) of four nucleotides or bases – (A) adenine, (G) guanine, (T) thymine, and (C) cytosine; these are arranged in a varying sequence. Each nucleotide consists of one of the four nitrogen-containing bases, together with a sugar (deoxyribose) and a phosphate group. The phosphates and sugars of adjacent nucleotides are linked together to form the backbone of the single stranded DNA polymerase chain. In the nucleus, DNA is double stranded to form a helix (the famed "double helix") as a result of pairing (bonding together) of complementary bases (so-called "base pairs"). It is the varying sequence of base pairs in the double helix of the DNA chain that provides the unique characteristics to the genes in each human, as well as the differences among species (9). Three nucleotides in sequence are called a codon, which codes for a single amino acid. Genes, which are essentially a series of codons, are sequences of nucleotides located at loci on the chromosomes and they are read in a series beginning from a fixed point on the chromosome. They may contain thousands of base-pairs of nucleotides. DNA is found in the nucleus of all cells including sperm, blood cells, cells in saliva, bone marrow cells and cells surrounding hair roots. Each cell nucleus contains all of the chromosomes; within a given human, the DNA is the same from cell to cell (10).

With the exception of identical twins, no two humans have identical DNA throughout their genome. The genome consists of all base pair sequences (roughly three billion) most of which in humans consist of non-coding regions interspersed with relatively few coding regions or genes. Therefore, if one were to compare the entire DNA of a person with a sample found at a crime scene, the accuracy of determination of similarity would approximate 100%, barring negligence or other error in the laboratory's process of assessment. But even DNA "fingerprinting" does not utilize the full range of information of DNA. (Until very recently the full genome was not delineated). Focused entirely on identification, "fingerprinting" is not concerned with the content of the genes, but rather relies on variations at specific regions at which samples of DNA are compared with one another to determine if there are similarities or if there is no match (11). Hence several markers are used to identify specific regions (or loci) on the genome. These loci are selected to reveal variability among individuals, marked by differences in the length or spaces between pre-specified loci.

A variety of genetic tests are available, but the main one used is single-locus and multi-locus restriction fragment-length polymorphism (RFLP) (12). PCR (Polymerase Chain Reaction) is also sometimes used when DNA is deteriorated or only small fragments are available. This test uses a process, which repeatedly duplicates and amplifies the small fragments to yield a DNA sample of sufficient size for analysis. This can be very useful for matching, but any contaminant that may be in the original sample also will be amplified, so that inaccurate or false positive results may occur. A newer test, Short Tandem Repeats (STRs) (repeating blocks of DNA) (13) uses PCR to amplify the segments of DNA to create a profile; a number of loci on the subject's DNA are examined, and the number of loci that match increases the probability that a sample came from a particular individual. It is said that if one locus matches, the chances are estimated at 1 in 500, whereas if all match the chance of inaccuracy in identification is said to be 1 in 82 billion. In the absence of an adequate nuclear DNA sample, mitochondrial DNA occasionally has been identified. Mitochondria within the cell are the main site for the metabolism that employs oxygen to produce high energy compounds used by the body. Mitochondrial DNA differs markedly from the DNA in the cell nucleus ordinarily used for matching studies. Nevertheless, it is inherited solely from the mother and is relatively unique in a given individual, so that on occasion it has been helpful for matching purposes.

RFLP allows comparison of DNA from an unknown source against known DNA. The DNA is extracted from a cell (e.g. from the nuclei of white blood cells in the blood) by chemical digestion and purified. The resultant DNA is then subjected to further digestion using restriction enzymes which act at specific sites cutting it into fragments of varying lengths. The fragments are then placed between two electrically charged poles and the resulting current of electricity causes separation of the fragments, a process called electrophoresis (14). The smaller fragments move more quickly through the gel than the larger, producing a pattern of fragments. The fragments are then transferred to a nylon membrane and the DNA strands are "unzipped" from one another at their base pairings. In order to visualize and quantify the bands, small DNA fragments tagged with radioactive probes are added to the nylon membrane where they attach to complementary base sequences. After washing off any excess probe material, the nylon membrane is positioned next to a sheet of x-ray film and exposed for several days. The result is a pattern of dark parallel bands on the film, referred to as an autoradiograph, or more commonly, an "autorad". The lengths of specific strands from similar loci in a reference sample are then compared to those of strands taken from the test sample (i.e., a remnant found at the crime scene) using the same loci; if they are closely similar the profiles are considered a match.

Should profiles not match at one or more loci, there is no match and excluding a laboratory error, one may conclude that the DNA is not from the same person. When there is indication of a match, population statistics (which differ among racial and other identifying groups) are used to assess the significance of the match (15). Even so, as we will see later, when a match occurs it is not a matter of irrefutable certitude (100% certainty); possibilities exist for false negative and false positive results.

THE USES OF DNA "FINGERPRINTS"

DNA fingerprinting has two basic applications in law enforcement (16). In one instance, other evidence points to one individual as a prime suspect, which warrants testing the suspect's DNA. Such an event is referred to as the "confirmation" case (17). Thus, testimony by the victim or an eyewitness identifying the accused, or a trail of blood or other bodily fluid, or other circumstantial evidence would prompt the police to secure a DNA sample from the suspect. At one time, obtaining the DNA sample might have meant drawing blood, but now a cheek swab will frequently provide the DNA. Also, it has been reported that a sufficient quantity of DNA can be obtained from a human fingerprint (18).

Evidence concerning the suspect's DNA and the DNA from a crime scene sample is usually presented in court as "match probability" (19) which quantifies the rareness of a random or chance match (20). Match probability can be stated either as the probability that the DNA of an individual from a given population will match that of the crime scene or that the population would match the DNA of a sample at the crime scene; both approaches are affected by the nature of the population and the number of loci examined.

Another use of DNA evidence depends on the presence of massive DNA databases developed in the fifty states from a class of convicted offenders and on the FBI's Combined DNA Information System (CODIS). When a crime has occurred and no specific suspect been designated, authorities can use these databases in what is called a DNA mining process or a "trawl case". For example, let us suppose that a 13-year-old girl is raped and murdered in a small community and although there is no obvious suspect, a crime sample (semen) available. In the "mining process" the police can compare the DNA of the crime sample against a database comprised of sexual psychopathic murderers including rapists, to determine if a comparable DNA profile can be found. The so-called "cold hit" occurs in the absence of any leads, with non-specific biological evidence which leads to identification of a suspect essentially unknown before the comparison (21). After the "cold hit", additional evidence would then be sought to confirm the identity of the perpetrator of the crime.

The DNA mining approach, which requires the existence of DNA from a class of known offenders in various crime categories with storage of the DNA profiles in an accessible database, has been in effect for some time in Great Britain; since 1995 DNA profiles of all persons charged with a "recordable offense" have been obtained routinely. It is anticipated that ultimately the database will include one-third of young English men (22). The FBI through CODIS has several hundred thousand profiles in databanks and it is reasonable to expect that it will become increasingly vital to law enforcement. Eventually the population of those subject to DNA tests will increase and may even broaden to include not only convicted offenders but also others not suspected of criminal activity (23).

Two reports issued by the National Research Council regarding DNA evidence present issues concerning the Trawl Case (24, 25). The National Research Council takes the position that when a match occurs between DNA evidence in a crime scene and a profile in a database, it should be seen as the basis for additional testing of markers at other loci (26), and that obtaining blood from the person is justified so that comparisons can be made with other loci not used in the first match. If the samples match a second time, the Council would allow that only the correspondence between the crime sample DNA and the suspect be admitted in court, since the additional correspondence could be legally construed as infringing upon the defendant's rights by introducing what may not be justifiable.

Should both the "screening" test and affirming test involving totally different loci match, the likelihood of this occurring on a randomly selected member of a defined population would be extraordinarily low--perhaps as unlikely as 1 in trillions. However, the Council would not allow this figure to be reported, although it would allow a finding that the second test came out affirmatively to be admitted. The lower probability is valid simply because both tests, when taken together, constitute at the very least a test with an expanded number of loci.

An initial match clearly indicates that the source of the crime scene sample was the suspect; it does not prove that the suspect actually committed the crime (27)--a conclusion that would require additional evidence. Lacking this, it might be determined that there is insufficient evidence to link the suspect to the crime, since it is conceivable that others may have one locus of correspondence with the DNA sample found at the crime. The larger the database the greater the chance of such a false positive. On the other hand, if there is only one match in a very large database the probability that the suspect was responsible for the crime sample is very high, and extending this further, if it were possible to include the entire suspect population in the database, then the existence of only one match establishes unequivocally (barring negligence in the collection and testing of the material), that the suspect is the source of the crime sample. Therefore, conclusiveness of proof of accurate selection depends on the nature and size of the population tested.

One final point of difference between the confirmatory case and trawl case relates to the significance of the affirming evidence. In the confirmatory case, the DNA fingerprint is taken after other evidence linking a suspect to a crime has been obtained; the preceding evidence is critical in profiling the suspect, and the DNA fingerprint confirms the significance of that evidence. In the trawl case the confirmatory evidence is obtained after the DNA fingerprint match has occurred, and there must be concern that the power of suggestion could affect the reliability of the conclusion regarding culpability. Thus, there is a danger that the confirming data may be skewed to comport with the results of the trawl DNA test (28), with the psychological effects of a positive DNA fingerprint being sufficiently compelling to offset an objective analysis of the supporting data.

CERTAINTY AND THE FRYE/DAUBERT AXIS

Under both the Frye and Daubert tests (29), DNA fingerprinting has been admitted in both criminal and civil cases. In some states, DNA testing is required by law in all persons arrested for a crime (30) and many other states have legislation allowing the results (as well as probabilities) to be admitted in court (31). Nevertheless, problems remain with the certainty of the science of DNA fingerprinting and evidence in general, so that the assumption of high reliability is not applicable in all cases. Thus, for the most part DNA testing is a precise science, but in the forensic setting a wide variety of difficulties can affect the results of a test (32). A critical step in DNA analysis is the application of digestive enzymes to break down the DNA. A given laboratory may vary its procedures, even though it has an established protocol. It may choose to use enzymes that are different from those used by most laboratories, or the enzymes may fail to work properly because of an inherent defect, thereby producing variant results, so that errors or deflections may occur in each step of the DNA test, distorting the final results.

In the Castro case in the late 1980s in New York (33), a sample of blood taken from a watch found at the scene of the murder was sent to Lifecodes Corporation for DNA testing. The results showed matching with the defendant's DNA and were reported to indicate odds of one in one million that a random match would occur. On further examination of the procedures and analysis, it became apparent that the laboratory had not used appropriate standards and the test was held to be unreliable.

DNA material undergoing testing may be contaminated or switched during the handling of the sample. Alternatively, during the collection or during DNA extraction processes, inadvertent admixing of the sample with other samples in the laboratory may occur. These errors could produce false positive results or inaccurate determination. Even in laboratories with effective quality assurance measures, errors or misidentifications can and do occur.

Ambiguities in testing provide another set of opportunities for error. It has been suggested that comparing autorads is generally considered a simple task (34). However, there are variations in the appearance of DNA prints. These can be caused by minor differences in the procedures, enzymes, samples, gel and other materials used in the testing, variations that can produce slightly different locations of bands, or even extra bands. For the most part these differences are overlooked if they are not major. Experts differ on the significance of extra bands, some taking the position that the presence of an extra band eliminates the possibility of a match; others have testified that an extra band does not affect the validity of results, and therefore does not exculpate a suspect (35). Adding to the possibility of error, a laboratory may be in contact with the prosecution and an exchange of information may occur, affecting the subjective judgment of the analyst of the DNA bands.

Finally, the statistical significance of the match may be exaggerated. In particular, where there are clear inconsistencies or ambiguities, these should be factored into any statement of probability or frequency. This is rarely done. For example experts have given figures such as a one in 4 billion chance of a random match in a particular test, discounting such factors as ambiguity created by minor differences in bands, or the presence of an extra fragment, that would affect the probabilities. Certainly, the presence of confounding factors should be considered in any conclusion on the likelihood of a match. This may not affect the admissibility of the DNA evidence into a trial, but it would undoubtedly have an impact on the jury's decision and therefore should be made available for their consideration. Taking into consideration the possibilities of contamination and ambiguities in interpretation, the probabilities may decline drastically from millions or billions to a one in X thousands chance that a random sampling would reveal a match.

In summary, DNA fingerprinting is generally a highly reliable process that results in statistically significant correlations between the DNA of a sample at a crime scene and a subject. This reliability applies not only in confirmatory case but also in the dragnet or trawling situation although exclusion is more definitive than confirmation in many instances. It must be recognized that the testing process is open to errors at various points along the trajectory toward establishing a match. Negligence, or the imposition of different standards can alter test results, which, along with ambiguities in interpreting result, can decrease the overall accuracy of DNA fingerprinting. Nevertheless, when compared with many other types of evidence admitted in trials, DNA fingerprinting is usually more reliable. The expanding use of DNA evidence in civil and criminal cases makes it imperative that standards for testing be applied uniformly in laboratories throughout the country.

NOTES

1. "DNA Said to Link a Friend to '75 Hoffa Disappearance". *New York Times*, September 8, 2001: page A-7.
2. See Howard Coleman & Eric Swenson: *DNA in the Courtroom: A Trial Watcher's Guide.* 3 (Dwight Halloway & Teresa Aulinskas eds, 1994)
3. A.J. Dodson: Comment: DNA "Line-Ups" based on a reasonable suspicion standard. 71 *U. Colo.L.Rev.* 221(Winter 2000).
4. For a discussion on DNA "dragnets" where an entire class of individuals in an area (such as males in a case of rape) are subject to DNA sampling while crime being investigated. The profiles of these men are compared against samples found at the crime scene, see: NOTE: Arresting crime: expanding the scope of DNA databases in America. 79 *Tex. L. Rev.* 921 (March, 2001).
5. Robert J. Goodwin & Jimmy Gurule, Criminal and scientific evidence 287 FF (1997).
6. See Dorothy Nelkin & Laurence Tancredi: *Dangerous diagnostics: the social power of biological information.* The University of Chicago Press, Chicago, 1994 pp. 151-154. Also, Goodwin & Gurule supra note 5.
7. Edward Connors et al., Convicted by juries, exonerated by science: case studies in the use of DNA evidence to establish innocence after trial 4, 7 (1996).
8. See Daniel J. Kelves & Leroy Hood eds. *The code of codes: scientific and social issues in the human genome project.* Cambridge: Harvard University Press (1992). Also, Muller-Hill, B. The shadow of genetic injustice. *Nature* 362:491 (1993).
9. James D. Watson et al: *Recombinant DNA* (2d ed. 1992) 14 ff, see also, Warren R. Webster, Jr.: Note: DNA Database Statutes & Privacy in the Information Age. 10 *Health Matrix* 119 (Winter, 2000).
10. Ian W. Evett & Bruce S. Weir: Interpreting DNA evidence: statistical genetics for forensic scientists 219 ff (1998)
11. Eric Lander: DNA fingerprinting on trial. 339 *Nature* 501 (1989)
12. Dorothy Nelkin & Laurence Tancredi, supra note 6 at 27ff.
13. See Goodman and Gurule, supra note 5. Also, Note, 79 Tex. *L. Rev.* 921 supra note 4.; Victor W. Weed & John W. Hicks: The unrealized potential of DNA. *Nat'l inst. of justice research action*, June 1998, at 7 ff.
14. See Kelves & Hood, supra note 8, at 380 ff.
15. D.A. Berry et al.: Statistical inference in crime investigation using deoxyribonucleic acid profiling. 41 *Applied Stat.* 499 (1992). See also Zollinger, Jay A.: Defense access to state-funded DNA experts: considerations of due process. 85 *Calif. L. Rev.* 1803 (December 1997).
16. Paul E. Tracy & Vincent Morgan: Criminology: Big brother and his science kit: DNA databases for 21st century crime control? 90 J. *Crim. L. and Criminology* 635 (Winter 2000).
17. Peter Donnelly & Richard D. Friedman: DNA database searches and the legal consumption of scientific evidence. 97 *Mich. L. Rev.* 931, 933.
18. See Richard Saltus: DNA in Fingerprints Used as Identifier, *Boston Globe* June 19, 1997 at A5.
19. Donnelly & Friedman, supra note 16 at 933-934
20. See Donnelly & Friedman, supra note 16. See also, Richard D. Friedman: Assessing evidence, 94 *Mich. L. Rev.* 1819 (1996)
21. See Tracy & Morgan, supra note 15, 635.
22. See Donnelly & Friedman, supra note 16, at 935
23. See Nicholas Wade: F.B.I. Set to Open Its DNA Database for Fighting Crime, *N.Y.Times*, October 12, 1998, at A-1
24. See Committee on DNA Forensic Science, National Research Council: DNA Technology in Forensic Science (1992). Also see Donnelly and Friedman, supra note 16, at 53.
25. Committee on DNA Forensic Science, National Research Council: The Evaluation of Forensic DNA Evidence (1996).
26. Committee on DNA Forensic Science, supra note 24 at 129.
27. See Donnelly and Friedman, supra not 16, at 938.
28. A. Tversky and D. Kahneman: *Judgment under uncertainty: Heuristics and biases. Science* 185:1124-1136 (1974).

29. See Brian Huseman: NOTE: *Taylor v. State*, Rule 706 and the DNA database: Future direction in DNA evidence. 22 *Okla. City. U. L. Rev.* 397 (1997), and Webster, supra note 9, at 123.
30. See Note: Arresting Crime: Expanding the scope of DNA databases in America, supra note 4, at 930.
31. Webster, supra note 9, at 123. See also: Andre A. Moenssens et. Al.: Scientific evidence in civil and criminal cases, 879ff. (4th ed. 1995).
32. See Jay A. Zollinger: Comment: Defense access to state-funded DNA experts: Considerations of due process. 85 *Calif. L. Rev.* 1803, 1808-1812.
33. See New York v. Castro 545 N.Y.S. 2d 985 (Sup.Ct. 1989)
34. See Zollinger, supra note 32, at 1810.
35. See Zollinger, supra note 32, at 1810.

CHAPTER 8

NOTES FROM THE TRIAL OF THE CENTURY

The Medico-Scientific and the Mythic

THE PEOPLE AND O.J. (1)

On June 12, 1994, at about 10:15 PM (according to the prosecution), or, between 10:35 PM and minimally at circa 10:50 PM (according to the defense), Nicole Brown Simpson and Ronald Goldman were murdered at her residence, 875 South Bundy Drive, in the Brentwood area of Los Angeles. The prosecution held that the struggle between victims and the assailant lasted one minute and fifteen seconds; defense experts argued for a scene, derived from "necessary" inferences from forensic evidence, that the struggle with Ron Goldman alone had to have lasted at least five, or fifteen, but most probably ten minutes (2). Such differences presented by experts for prosecution and defense were not trivial since the opportunity space and time for targeted defendant O.J. Simpson to have committed these complex violent crimes would play a huge role in deliberations about whether there was sufficient time for him to have executed the deeds, returned to his home on Rockingham (a six or eight minute drive), disposed of all the clothes and other bloodied evidence, tidied up, and met the limo driver who testified he first saw Mr. Simpson at 10:56 PM. Opening statements began on January 24, 1995, and the trial ended on October 2, 1995 with a jury verdict of acquittal for Mr. Simpson on all charges.

One of us (Lola Romanucci-Ross) watched, studied, and indeed researched the trial (as a cultural and medical anthropologist), observing Court TV when time permitted, taping all other sessions when it did not, and viewing and taking notes from the tapes on weekends and evenings. This was supplemented by daily newspaper accounts, TV news and talk shows, whenever possible, to determine if the reportage accurately reflected what had happened in the courtroom that day. The focus was on the trial; the purpose was to observe and analyze the uses of scientific evidence and the rules for its admissibility and argumentation in a court of law. It was not to determine guilt or innocence of the defendant (although the news media kept score of winners and losers on a daily basis). Even though media pundits were undaunted, this was a difficult task indeed, for the prosecution would present a convincing argument only to incite the defense to an equally convincing refutation. Subsequently, books by various participants in the trial were acquired and read for differing perspectives, "explanations", apologias and phenomenological views. Presented here is a summary of certain salient events, with emphasis on the unfolding of contentions about the scientific evidence that propelled the trial, and interpretations of the evidence that appeared to have influenced the jury's decision.

This trial will retain its intellectual interest, long after the allure of media frenzies and personal displays of outrage have faded away, for it has major relevance in the context of scientific evidence and the law in debates about the methods of scientific inquiry, the logics of discovery and, of course, all the extraneous factors that often come into play in a trial, for example, the element of "chance". It also provides a lesson in culture change or, more likely, mirages of culture change, with regard to gender roles and racial attitudes, stereotypes of government agencies, and a host of other social and economic pressures. The prosecution flooded the media with promises of incriminating evidence (a "mountain of evidence"), most of it DNA evidence which, we were told, would convince everyone of the defendant's guilt. In addition, media persons (in newspapers, on television, radio, etc) repeated these assertions daily, usually adding or implying that DNA evidence, being irrefutable, would prove Simpson's guilt. This was a large initial step forward for the prosecution, for what had to be overcome in the "court of public opinion" (the public, of course, was the pool from which jurors would be chosen) was the esteem and affection that America had for one of its favorite football heroes, O.J. Simpson.

The anthropologist seriously committed to following this trial in all its aspects and searching for an understanding of all that transpired (even as members of the jury had sworn to do), looked forward (with notebook in hand and recorder on desk) to vistas of information from the "mountain of evidence" promised by District Attorney Gil Garcetti and his lead prosecutors Marcia Clark and Christopher Darden (3). Central to their argument was the blood evidence, its loci at the crime scene(s), and the DNA it contained.

THE MEDICO-SCIENTIFIC: DNA

Certainly, the bloody event on the night of June 12 at the Bundy crime scene had provided more than enough blood samples. These were collected on the morning of June 13, by wetting small swatches of fabric which when applied to blood stains, moistened the stains so that the blood could be absorbed by the swatch. These were placed in plastic bags and set aside without refrigeration, and stored in a hot van for twelve hours (4). They had not been dried as soon as possible after collection (the usual practice), which allowed bacteria to attack the DNA fragments. Because of this, much of the original DNA had degraded beyond retrieval by the usual testing procedures, so that PCR (polymerase chain reaction), then the latest DNA technology, had to be used. However, as indicated in the preceding chapter, the PCR method, by replicating very small fragments over and over, introduces the risk of amplifying contaminants along with the pertinent DNA, thereby offering the possibility of false positive results. Dr. Kary Mullis, Nobel Laureate and the most prominent figure in the development of the PCR method of DNA analysis, although he was not called to testify, had this to say when interviewed about the PCR method in this context:

"The DNA tests used in the Simpson case are not suitable for forensic purposes.

They are not suitable for drawing any scientific conclusion... furthermore, the methods of the LAPD Laboratory and the California State Laboratory at Berkeley are seriously compromised. Basic scientific principles are being so violated as to leave their conclusions worthless" (5).

Prosecution experts did not dispute the contention that DNA rapidly degrades in the conditions under which the samples had been preserved, nor did they dispute that "subsequent contamination of such a sample by a second person's DNA can cause it to falsely match the second person on a DNA match"(6). The context for this admission by the prosecution had taken place when a young LAPD laboratory technician Colin Yamauchi took the stand for direct examination by defense lawyer Barry Scheck, who had studied Yamauchi's laboratory notes on procedures. Yamauchi had admitted that he had spilled some of the blood from Simpson's reference vial (blood drawn directly from him by the police), and on direct questioning recalled for the first time that it leaked through a chem-wipe onto his laboratory glove while he was in the LAPD laboratory handling all three vials, two containing blood from the victims and one from Simpson (7). He testified that the blood on his glove might have reached the worktable as well, and might thereby also have transferred some of Simpson's blood to the famous "bloody glove" from Rockingham. Furthermore, the swatches from the Bundy crime scene may have been similarly contaminated. He was not sure when he might have changed his gloves and indicated that he did not change the working surface very often. In response to a question he confirmed that once blood did spurt out of Simpson's reference sample vial (8).

For the defense, this episode in the trial was the prologue to a partial explanation of why Simpson's DNA was found wherever Ron and Nicole's DNA appeared in mixtures. It was also the opening for Dr. John Gerdes, a microbiologist who ran a DNA laboratory and who had inspected the LAPD (Los Angeles Police Department) laboratory a year earlier, to be called to the stand to describe the LAPD laboratory as "a cesspool of contamination". Among the unacceptable risk factors cited, the habit of handling the blood reference samples from the two victims and Simpson's blood sample at the same time was a continuing one. Also, degraded samples and reference samples were handled at the same time. There appeared to be little or no changing of paper on the work table, or changing of gloves by those who worked there, and reagents were not changed in a regular manner. If one could find these testimonies credible, then it would come as no surprise that the swatch from the Bundy crime scene with the largest quantity of Simpson's DNA (swatch number 51) was the first swatch the laboratory technician touched to work with right after handling the Simpson reference vial. The vial already had some blood on the outside, "some smeared, some caked". The second swatch he touched had the next highest, the third the third, and so on down to the last swatch, with the amount of Simpson's DNA on the swatches directly related to the sequence in which Yamauchi handled them (9). It would be difficult to explain this correlation in any other way. It was, in fact, Dr. Gerdes' reports of his inspection that had prompted Barry Scheck to study scrupulously all of Yamauchi's notes in the discovery material as he tried to follow closely the sequence of events in the laboratory on the day after the murder and before the testing began. He was able to find a pattern to support the contamination hypothesis, and to predict what would be found based upon his hypothesis.

Blood samples taken from the crime scene, i.e. those collected and "preserved" on swatches, had never been counted by the criminalists who collected them, Andrea Mazzola and Dennis Fung. The former was on her first collecting assignment in an "important" case. Neither were the blood sample swatches counted when placed in tubes for drying, or when taken out of the tubes. There was no documented booking of samples until June 16, 1994. Dates, or lack of them, for blood samples collected turned out to have considerable significance in the case. The late "blood finds" (e.g. the one on the back gate at the Bundy crime scene) were made on July 3, three weeks after the murders. Photographs of the gate on dates immediately after the murders, and several days after that, showed no blood. The Simpson "bloody socks" beside his bed, on the floor, were found only on August 4. Earlier videos of the bedroom, including the carpet area beside the bed show no socks. Both these (later) finds had high concentrations of DNA, higher than in the blood collected at Bundy soon after the crime was committed. Both also contained high levels of EDTA, the anti-coagulant preservative used in the so-called "purple top tubes" in which reference blood samples are stored. This was crucial to the defense argument that all the laboratory results may agree, strengthening the conclusions of each other, but if you don't know how the blood got there, they will they will tell you nothing about who was there and what happened at the crime scene. As a trial participant later reflected, "a number of scientific issues in the case had little to do with the fundamental science underlying DNA or the details of laboratory procedures for typing DNA, and everything to do with the potential for cross-contamination of samples before they reach DNA laboratories" (10).

Dr. Henry Lee had informed the defense, when they sought his expert testimony, that his sole interest lay in presenting the scientific evidence as he found and interpreted it, regardless of which side of the dispute it might favor. At one point he noted, "lots of blood everywhere, should be a lot of blood on the murderer... if collected evidence properly, real killer would be identified. They did not do their job from day one" (11). Barry Scheck, one of the defense lawyers, had devoted much effort to DNA litigation. Although early in his career he had challenged DNA admissibility, he later became a strong advocate for its use in excluding a subject as a suspect. Both he and his associate Peter Neufeld were aware that, although DNA evidence is almost perfect for "excluding", there are a number of limitations to the use of DNA for the exact identification of the person who actually did commit a crime. (This conclusion provided the impetus for their Innocence Project, which has cleared an impressive number of prisoners through DNA studies by proving they were not the perpetrators of crimes for which they had been incarcerated). In the Simpson case, they decided that the prosecution's "mountain of evidence" was a molehill primarily because the blood evidence and the DNA evidence was "compromised, contaminated, and corrupted" (12).

CREATIVE FORENSICS AND THE DNA EVIDENCE

DNA testing, when all the applicable scientific rules are scrupulously followed, may well tell a true story about whose blood was found at a crime scene; however, it can never tell you how the blood got there. The LAPD (Detective Vannater) was in possession of Simpson's reference blood sample, which was drawn while Vannater was questioning Simpson, a procedure for which Simpson volunteered. This occurred after an interrogation by Vannater adumbrating that Simpson "naturally" would be considered a suspect:

"Understand, the reason we're talking to you is because you are the ex-husband...".

"I know, I'm the number one target and now you tell me I've got blood all over the place". (13)

Having fingered the prime suspect, the prosecution waited three months before samples of Simpson's blood were sent for testing to several laboratories (14). Among events that may have prolonged the waiting period: "blood evidence" of Nicole Brown Simpson and Ron Goldman were found on the console of Simpson's Bronco after weeks of unsupervised access to the vehicle by many persons. Notably during the trial several persons unknown to each other testified that they had been in the Bronco some days after the crime with the expressed goal of seeing "all the blood" they had read about in newspapers or heard about on TV news reports. All these individuals testified that they had observed no blood, although Marcia Clark tried unsuccessfully to get each one to admit that he or she suffered from poor vision. One of these witnesses, an LAPD detective, testified she saw no blood in the Bronco, and that she had entered it expressly for the purpose of inspecting the vehicle thoroughly for evidence of this sort. She also testified there had been no log kept of persons entering and exiting the vehicle. Many individuals had had access, and one had taken a receipt for Nicole's dry cleaning from the glove compartment as a souvenir. One witness, extremely irritated by the misleading newspaper accounts of "blood in the Bronco" left his fingerprints on the console to prove he had been there and discovered the inaccuracy of these accounts.

Several LAPD detectives, led by Mark Fuhrman, immediately reported on the night of the murder that the Simpson residence was also a crime scene. From that point on, detectives and others moved constantly between the Bundy and Rockingham crime scenes, without apparent concern that the detectives at the Bundy scene walked around gathering blood, hairs and fibers on their shoes while the bodies of the victims, covered with blankets taken from Nicole's bedroom, were being moved. When description of these events would be denied by a prosecution witness, the defense would show slides of the detectives or other witnesses stepping into a pool of blood or on other transferable items such as hair and fibers.

The chain of custody of evidence was severely compromised by the early failure to document how evidence was collected, and because those first swatches were neither numbered nor counted, the number of swatches on which the technicians and analysts began their work was not knowable. Later, some information about the swatches came to light, and that is how Dr. Henry Lee was able to target them to prove his conclusion that "something is wrong here".

Dr. Lee found that some swatches were in bindles that had not been initialed; Mazzola had testified, however, that all bindles from the crime scene had been collected and initialed by her. Examination of the non-initialed bindles led defense experts to argue that they were substituted swatches, which presented the DNA that the LAPD wanted them to have. Had these bindles been initialed, this conjecture would never have led to subsequent investigation which indicated that blood on the non-initialed swatches had, in some instances, 135 times and in other instances 270 times, more DNA than other blood drops collected from the same scene. Also, the EDTA content of these samples suggested an origin, not from a crime scene, but rather from a purple-top tube. Henry Lee and Barry Scheck found evidence of "wet transfers" in some of the bindles, although the testimony of criminalists Mazzola and Fung had been that those samples were left to dry for fourteen hours. Evidence and testimony suggested that these swatches were placed there "wet" to replace the original swatches collected from the crime scene.

At this point in the trial, a devout trial observer and note-taker would recall prior witness testimony that criminalists who worked at the LAPD laboratory underwent little or no training the in collection or preservation of DNA.

MISSING BLOOD

The defense offered an explanation for the blood that "went missing" from the vial of the Simpson reference sample; it was that identified as Simpson's blood in the "later discoveries" on the socks in the bedroom, on the back fence, in the Bronco, etc. At the very least they put it to the jury as an amazing coincidence, for certain events begged for explanation and/or a rationale: First, in the preliminary hearings (15), Thano Peratis, the nurse who drew blood from Simpson on June 13 as he was being interrogated by Vannater, testified that he labeled the amount drawn as 8 ml. He did so on this occasion, as he had on countless previous occasions, by looking carefully and ascertaining that it was "precisely" between 7.9 ml and 8.1 ml. However, Barry Scheck had noted that the amount of blood in the tube was documented as 6.5 ml by the first technician in the LAPD laboratory who began an analysis of the Simpson reference sample. That this was 1.5 ml less than the amount recorded by Peratis was, of course, unknown by the laboratory analyst. From time to time, prosecution witnesses were presented to explain the missing blood, but with what appeared to be with little success. Toward the end of the trial, the prosecution introduced a new strategic approach. Prosecution lawyer Hank Goldberg was given the unenviable task of visiting and tape recording a post-operative, enfeebled Thano Peratis to establish a revisionary view of what he recalled about labeling the amount of blood on June 13 of the previous year. He remembered that he had made a mistake in noting the amount collected, now recalling that it was 6.5 ml. This event was rather embarrassing to most who watched the videotape during the trial. Later it raised reasonable doubts for some jury members, who said, after the trial, that the missing blood and presence of EDTA in blood drops and blood stains discovered "later" constituted the "turn around point" for them in what they had initially felt was the prosecution's very good case against Simpson (16).

What might have caused this attrition of Simpson's reference blood sample? After the blood was drawn at the close of his initial interrogation of Simpson, Detective Vannater did not go directly to the nearby Parker Center to book the blood, as he had always done in his 26 years with the LAPD. Rather, he took the blood vial with him, carrying it on his person for almost three hours from 2:30 PM to almost 5:30 PM on June 13, as he went back and forth from the Bundy and Rockingham crime scene, booking the sample later on June 16. During this period Vannater had access to three reference samples – those of Simpson, Nicole Brown Simpson, and Ron Goldman (17). The stain on the sock from the Simpson bedroom was composed of the blood of Nicole and Ron. Defense expert Leland McDonnell testified that the blood had been pressed on the sock fabric, not spattered, as it would have been in a battle for survival, and that it was compressed through and transferred to the inside of the opposite side of the sock, which could not have occurred if there had been a leg and a foot in the sock. McDonnell presented much more evidence for his conclusions, including the presence of EDTA in that stain, suggesting an origin from blood samples. Criminalist Dennis Fung had documented that he had found no blood on the socks when he collected them from Simpson's bedroom, and consequently he had not put them into a container reserved for bloody clothing. Michelle Kestler, the LAPD laboratory supervisor wrote "no blood found" when the socks came to her (18). On June 29 Judge Ito ordered an inventory to determine what blood evidence was available that might be shared with the Defense. Once again, there was a report of the socks with "no blood". Furthermore, a video for insurance purposes taken on June 13 at Rockingham showed no socks in the Simpson bedroom.

Three bloodstains were collected from the back gate of Nicole's residence at Bundy on July 3, 1994, about three weeks after the murders. They had been missed or overlooked by several police officers, detectives, and criminalists at various times, in the explanation offered by the prosecution. A photograph taken on June 13, one day after the murders, showed no blood on the back gate yet the stains obtained on July 3 were the largest collected; the blood was not degraded as were the blood drops collected in front of the Bundy house, and it contained high concentrations of DNA as well as the presence of EDTA preservative (19).

AND OTHER PERPLEXITIES IN THE DNA DATA

The problems with DNA in this trial were many, including useless samples due to poor storage conditions, contamination, false positives, presentations of data that often relied on the meanings of "trace" and "hint", many improperly packaged specimens, samples containing other (non-human) DNA from the crime scene, some laboratory errors, and "band shifting" in processing (which may be caused by deficient electrophoretic gel, or overloading of one lane). It is difficult to overstate the importance of such factors, which remained unrefuted in the trial. Nevertheless, as we shall see, public belief in the infallibility of DNA evidence and unquestioning trust in what is perceived as "science" had much to do with what will be explored later in discussing the trial and its aftermath as a cultural phenomenon.

JOUSTING WITH NUMBERS

Statements on statistical significance and their validity are often controversial. As an example, there are controversies about how to reckon frequencies in mixed blood from unknown sources. In forensics, an expert might determine the frequency of each marker of an allele and then multiply the frequencies together; other experts believe that this distorts the results, and along with some other experts, contend that that frequencies of each marker should be added. Several points of contention arose between Bruce Weir, expert for the prosecution, and the genetics and biostatistics experts for the defense. Each allele has a marker; however, in the DNA DQ-alpha system of testing (relied upon by Weir), not all alleles have separate markers and therefore in a mixed blood sample (from two or more individuals) some alleles are masked and therefore inferred by the presence of other data. Therefore the presence or absence of Simpson's 1.2 allele could not be reliably determined in some mixed samples (20). But Weir testified that he simply assumed that the Simpson allele was present; underestimating the frequency with which this would occur in control samples colored his results. Confronted by the defense, Weir did not dispute this introduction of bias against the defendant Simpson, but replied, "As it turns out, it looks that way, yes" (21).

Also, relevant to interpretation of the evidence were: size of the database and, just as importantly, the representation of Whites, Blacks and Hispanics in the data base. As anyone who has done field research in Latin American countries knows (22), a Hispanic can be preponderantly or more -Caucasian, or Amerindian, or a mixture along a continuum of earlier discrete populations (*mestizo*); many African-Americans in the United States include Caucasian ancestry in varying degrees and in some cases, Amerindian or Hispanic as well. Some cities show large variations in data bases, e.g., an allele might have a one percent frequency in a specific subgroup, while the frequency for the homozygote for the allele would be calculated to be one in ten thousand for the large group, but one in twenty for the subgroup. Must not this sort of information be firm if not indisputable for frequency estimates to be valid? Should the relative possibility that the match is incorrect be factored into the probability of identification?

Do likelihood ratios include probabilities of laboratory errors and coincidental matching? What does one in a trillion really mean, since the suspect need not and should not be made to appear that he is competing with unknown populations, and populations yet unborn, as having been "likely" to have committed a crime? Concerning error and exaggeration about DNA evidence in trials, Koehler wrote that although it is not very likely that that one randomly selected person would have the same DNA profile with another person (or suspect), it may be quite likely that others do share this profile. "Even for a frequency as small as one in one trillion, the chance that there are others who share the profile is greater than one in two hundred" (23). Relevant to this, the blood drops leading away from the victims on the patio at

Bundy, those degraded drops presented by the prosecution as belonging to Simpson (which should have been presented as not excluding Simpson) were discussed by LAPD blood expert Gregory Matthieson. In response to defense questioning) he testified that about 43% of people in Los Angeles had the blood characteristics found in the "trail of blood drops' in the patio at Bundy.

Koehler avers that likelihood ratios or the ratio of two coincidental probabilities are not easily understood, not only by jurors but the officers of the court as well, in many instances. Yet much time and fury was spent on such topics in this trial, and in the sidebars. This area must have been a low point for the jury, considering the reasons that a number of jury members gave for their verdict; whatever the reasons for which the jury unanimously chose acquittal, the statistics of the probabilities are nowhere to be found in their post-trial recorded comments (24).

FROM SCIENCE TO "SPECIALTIES" AND OTHER INFERENTIAL ARTS

Star witness for the prosecution, during cross-examination by F. Lee Bailey, Fuhrman demonstrated that a policeman would lie under oath about a fact, to help the prosecution get a conviction. Fuhrman did so and was later called back to testify about his taped racial slurs on African-Americans; his remarks on this topic on the tapes were accompanied by comments on what a policeman could and should do to rid our society of these people. During recall, the defense took the opportunity to query him about his personally discovered evidence in the case against Simpson i.e. discoveries made by him after he had been officially removed from the case. Fuhrman asserted his Fifth Amendment privilege on all counts (the jury did not learn about his "taking the Fifth" before reaching their verdict).

Fuhrman had been among the first few detectives to arrive at the Bundy crime scene. Called at 1:05 AM on June 13, he was notified at 2:30 AM that Robbery and Homicide Division detectives were now in charge of the investigation, and that he was dismissed from this assignment. Yet he managed to be the lone discoverer of several crucial items of evidence central to the prosecution case (25). Among these was the "bloody glove" he said he found on Simpson's property during a warrantless search, which occurred after he had convinced Vannater and one or two other detectives that there might be "bodies or endangered persons" there. This glove was the mate of one found at the Bundy crime scene, both of which were too small for Simpson's hands during a courtroom "demonstration". Since the prosecution argued that they once fit but had shrunk because they had been bloodied, arguments followed over the nature of leather and shrinkage phenomenon, and whether such considerations would be admissible scientific evidence.

Expert witness for the defense, Herbert MacDonnell, conducted a shrinkage experiment using Isotoner leather gloves having all the attributes of those at the crime scene. Using his own blood to soak the gloves and then placing them under similar ambient conditions to those of June 12, 1994, he demonstrated that they do not shrink when allowed to dry. But the glove research continued. Richard Rubin, a former Isotoner executive, was brought by the prosecution from New York as a glove expert to testify about glove sizes, types of stitching, and how many pairs of such Isotoner gloves had been sold in New York and Washington. However, no evidence was presented that Simpson had ever bought or owned such gloves.

There were other problems with the bloody gloves beyond their lack of fit. The glove was still very wet with blood when reportedly found by Fuhrman at Simpson's Rockingham residence and all experts in the trial agreed that it should have been dry hours earlier had it been left by Simpson since he departed for the airport just before 11 PM on June 12. In addition, Fuhrman had requested that he be photographed by Rokhar, an LAPD photographer, while posed pointing to the bloody glove he had found at Rockingham and brought back to the Bundy scene at a time he asserted to be 7 AM on June 13. However, it was later demonstrated by the defense that the photograph was actually taken before 4:40 AM, while it was still dark, a time well before Fuhrman claims he found it at Simpson's Rockingham residence.

Douglas Deedrick, FBI agent and expert on hair and fiber evidence was asked to assess a "rare" beige fiber found at Bundy which resembled those of the carpet in Simpson's Bronco. Also, hair found in a cap at Bundy was not inconsistent with Simpson's. In cross-examination of Deedrick, defense lawyer Robert Blasier led him to admit that hair and fiber science was based on inference and subjectivity, that hair was not unique to a person, and that no one could say that a hair sample is a "match". Indeed Simpson was a frequent visitor at Bundy, often drove his children there from places he had taken them in his Bronco. Moreover, it was conceded that the only conclusion to be made about the fiber was that someone who had access to the Bronco had been at Bundy at some time.

AND OTHER MINOR ARTS OF PERSUASION

Developing strategies to control outcomes in the trail was, understandably, the goal of both sides in this litigation. Creativity in forensics is not unlike creativity in art or music or flower arranging; what is left out is just as important as what is included. The defense was bent on certain events not being included, since the lawyers felt these were swords that could cut both ways. Some events were not mentioned or even alluded to so that those doors would not be opened. This did not seem to create a problem since the judge almost invariably denied defense motions, by count at least two-thirds of their motions and requests were denied. The prosecution had many more management opportunities, and were successful in keeping certain evidentiary items, desired for inclusion by the defense, from entering the trial.

Examples of inadmissibility:

Nicole's mother's glasses were found at the murder scene, next to Ron's body. She inadvertently had left them at the Mezzaluna Restaurant and Nicole had asked Ron Goldman to bring them to her house. This was important evidence because there were bloody fingerprints and blood smears on the lenses, and the envelope containing them had been opened, handled and folded while the blood on it was still wet; the jury and the public were never told anything about the fingerprints. When found, there were two lenses in the envelope (as noted by Baden on June 27), but after being stored there was only one lens (noted on February 16), nor was there a record of the envelope originally containing two lenses. Evidence on the lenses was also never followed up by the LAPD, or if it was, there is no record of it. If there were results, positive or negative for either side, they were never shared. DNA associated with blood type B was found under Nicole's fingernails; this was not a characteristic of blood from Nicole, Ronald Goldman, or Simpson.

Possible connection to drug-related relationships of Nicole and particularly through Nicole's close friend Faye Resnick, who wrote of their drug dependencies in her book (26), might have been considered. On the night of the murders Resnick called Nicole from the high-security detoxification center to which she had herself admitted a very short time before. In this connection, the defense counsel wanted Brentwood chiropractor Christian Reichardt to testify concerning, for example, his knowledge that Faye frequently went to Nicole's place to freebase (crack)... "a thousand dollar a day cocaine habit" for which he "kicked her out of the house several times", and slept with a shotgun near his bed fearing "visitors" coming to collect on Faye's debt (27). This testimony was disallowed by the Judge, who had admitted Ron Shipp's recounting of a dream he claimed Simpson had described to him. (Simpson stated, out of court, that Ron Shipp was merely an acquaintance who "hung around" with groups who occasionally dropped in at Rockingham to drink or chat or swim in his pool, and was never a friend.)

Among these examples of omission, we should include fingerprint evidence, about which more will be discussed in general later. At the Bundy crime scene seventeen fingerprints were discovered, eight of which were identifiable. None of these was Simpson's, although the killer had left behind a pair of gloves, one of which was found at Rockingham by Fuhrman. Nine fingerprints were not identifiable i.e. did not belong to any known person who might have been there for any reason. This was never followed up, nor were any searches done for identification of these fingerprints. Or, if this was done, the results were not revealed.

OF THE TIMELINE, AND HUMAN ANATOMY/PHYSIOLOGY

The time needed to commit these acts of mayhem and murder, to get away from the scene into another frame of behavior, to eliminate all the evidence such as knife or knives, bloody clothing, etc, was of paramount importance to both sides. F. Lee Bailey had stated early in the trial the timeline alone would exonerate Simpson. However, he had not reckoned with the ability of scientific experts to design choreographies for reconstructing events of the killings (resistance behavior of victims, responses of killers, the nature of the terrain), as well as their skills in theorizing about the nature of organ responses to various types of injury. Nor was it expected that the narratives of the witnesses called to testify about their experiences near the time of the murders would prove critical in the litigation.

The prosecution sought to prove that the murders were committed very early that evening, as early as any bit of evidence would lend support to their position. The purpose of establishing an early event was to accommodate all the episodes that would have had to occur. The earliest time they could propose was 10:15 PM. This was accomplished by presenting witness Pablo Fenjves, a writer, who testified that he heard the wail of a dog at 10:20 PM somewhere in the neighborhood. After skillful prodding by the prosecution on direct examination, he conceded that it might have been 10:15 PM. A second witness called by the prosecution, Steven Schwab, testified that he was accosted by an Akita (the breed of the Simpson dog then spending most of his time at Bundy); it had followed him home and he later discovered it to have bloody paws. At first he testified that the time was after 11 PM, but after direct questioning by the prosecution, he thought it might have been closer to 10:55 PM.

Witnesses not called by the prosecution but summoned by the defense included Ellen Aaronson and Danny Mandel, who testified that they were through with their unsuccessful "blind date" and passing by the gate to the Bundy crime scene at 10:28 or 10:30 PM; they both were aware of time (both rather eager to end the evening). They neither heard nor saw anything; it was a quiet pleasant June night. The sidewalk they were on had a full well-lit view of that small area, which, from the same spot outside the gate, sometime after midnight Sukru Boztepe and Bettina Rassmusen saw the bodies. (They had wanted to rid themselves of the Akita which had been "stored" with them by Steven Schwab, and so they decided to take the Akita for a walk and follow the dog to its home). Francesca Harmon drove by that address at 10:25 PM, heard and saw nothing. Denise Pilonak, who lived next door to Nicole and who came out of her house to bid goodnight to a visiting girlfriend, also testified that she had initially heard and seen nothing; however, at 10:35 PM she had heard a dog barking, and this continued for about 35 minutes. In response to rigorous grilling by a visibly upset Marcia Clark (her timeline now problematic as recounted by a truly proximate neighbor), Denise stated that she was certain of the time because, after her friend left, she had called her mother; the time and duration of that conversation was recorded on her phone bill.

The defense then called Robert Heidstra, a car detailer who lived in a small apartment in the neighborhood. He testified that he, with his two aged dogs, were on their nightly stroll, and stopped in the alley no more than 75 feet from the Bundy crime scene. At 10:35 PM he heard a dog barking and two men arguing, one of whom shouted, "hey, hey, hey". He also saw a white Chevy Blazer drive away, in a direction opposite to that of Simpson's Rockingham residence. Mr. Heidstra had called the District Attorney's office to offer this information, but Christopher Darden, who received his message, had chosen not to call him to testify. Called by the defense to tell his story, Mr. Heidstra was then subjected to a "cross" that most would have found demeaning; his small and basement-like living quarters and his low status as a mere car detailer were emphasized with the purpose (it seemed) was to show that a person such as this would not be there to tell the truth but to curry favor of the wealthy defendant. This was explicitly said in court by Darden, who then asked him if the car could have been other than a white Chevy blazer. After several negative replies, and stating that his work well qualified him for his assessment of the vehicle, Heidstra finally gave in to the logic that, of course it "could have been" anything, and when further pushed on this, yes, even a Bronco, but what he *saw* was a white Chevy Blazer. (The next day, one could read any newspaper, turn on any TV station, and learn that a car detailer testified that the car he saw leaving the crime scene could have been a Ford Bronco).

Darden would later tell the world in his book that he and Marcia should not have tied themselves down to the idea that the murders happened at 10:15, for, as far as he was concerned, "the murders could have occurred as late as 10:40 PM and Simpson would still have had time to drive for 5 minutes back to his estate" (28). Of course the time to Rockingham as clocked by the police was between 6 and 8 minutes, which would have put him at his residence at 10:47 PM, and the limo driver Alan Park testified that he saw Simpson dressed, packed, and all ready to leave for the airport at 10:56 PM, describing his demeanor as normal, pleasant. In 9 minutes, Simpson would have had to dispose of the bloody clothes and knife, thoroughly clean out the Bronco which would have been covered with blood (according to all experts assessing the scene), jumped over his fence, run into his own air conditioner where he dropped one bloody glove (according to Detective Fuhrman), cleaned up the rugs on which he had walked with the famous Bruno Magli shoes, shower, dress, and get down the stairs to greet the limo driver. (It was interesting to note that in the subsequent civil trial, that Prosecuting attorney chose not to argue the timeline at all.)

Although challenges about inferences from medico-scientific evidence in this trial featured DNA analysis, there were other challenges, such as the meaning of quantities of EDTA in the blood, as mentioned earlier. Another aspect, relevant to the current discussion, was the time it took victims Ron Goldman and Nicole to die from the assaults. Experts on both sides had characterized the events of this hideous crime through analytic interpretations of hemodynamic events to determine this time interval. As noted above, Werner Spitz testified for the prosecution that the actual killing time was one minute and fifteen seconds. Henry Lee and Michael Baden testified that it had to take at least five minutes and, from all the evidence, probably took ten minutes. The evidence they relied upon: the thirty cuts on Ron Goldman's open hands, "vertical" drops of his blood in different areas in front of the house suggesting a moving battle, with blood spilled during stationery periods; footprints, some deeply planted in several spots which, along with the vertical blood drops and the finding that his beeper and keys had fallen out of his pockets at different locations, indicated that Ron Goldman had struggled on his feet for a considerable period of time. Forensic evidence suggested that wounds to the neck came first and that the last wounds were in the chest. Multiple stab wounds were also found on legs, abdomen, scalp, and arms. Nicole also died of multiple sharp wounds, some on the hands, compatible with defensive wounds, but the fatal stab was the deep incised wound of the neck, from left to right (29). Reconciling autopsy information with the timeline problem to reach the desired conclusion fell to Werner Spitz for the prosecution, and to Henry Lee and Michael Baden for the defense. The jury later stated that they were impressed with Henry Lee in this case because he took into account, not only the nature of the injuries and Baden's account of physiological response under such circumstances, but added his detailed scrutiny of the disturbances on the ground, surrounding bushes and fence, a possible footprint mark on Goldman's sweatshirt (not from a Bruno Magli shoe) as well as evidence concerning the envelope containing Juditha Brown's glasses featuring bloody fingerprints not linkable to any known person. After the trial, a number of the sequestered jury members, did try to address for interested readers, the logic behind their deliberations about the items of evidence that had influenced their final decision (30).

AN AESTHETICS OF MOTIVE

That people generally receive data or information and interpret it only in a manner that makes sense to them appears to be the case across cultures and through time, and the common format is the narrative. What is needed, it seems, is a "story-line", and the prosecuting attorneys in this case, the side that had the burden of proof, made major use of this instrument to explain the evidence and the motives of the characters in this painful drama. The defense, of course, only had the opportunity to poke holes in the story line in an attempt to destroy the fabric of evidence and motive. Judge Ito rarely granted the defense the opportunity space to present their own narratives – of drug use, of connections to suppliers, of witnesses who could have elucidated past relationships (31), of letters written by Nicole to Simpson's daughter, and a number of other similar exculpatory possibilities for Simpson.

The jury had been given strict instructions, repeated daily, as to what their role was to be; the public, not being assigned a role, became the target for information that would influence public opinion. The LAPD distributed a 911 tape, of some years earlier, of Nicole calling the police, with Simpson screaming in the background. (To this day almost anyone who did not follow the trial will say that this provided proof that he used to beat her, but the tape was clearly recording his anger at her behavior in their home with a current lover while the children were sleeping in their bedrooms at the top of the stairs. One had only to listen to the entire tape to ascertain this, but only fragments of the tape were usually played for the public). Also (on the tape) she told the LAPD contact on the phone that he was leaving, that he had not harmed her and that everything was all right. Before and during the trial, the LAPD and the D.A.'s office frequently released other materials about incriminating results of DNA tests that had not yet been done, and evidentiary revelations that turned out to be untrue, such as the discovery of a bloody ski mask in Simpson's house and incriminating findings in his golf bags (32).

POST-TRIAL TRYING TIMES FOR JURORS

The jury was given a task, or a trust: the twelve members were to consider only the facts in the case, facts that were so declared in court. They were not to be influenced by pity for one side and/or contempt for the other. Each fact had to pass the reasonable doubt test, and the possibility of innocence had to be considered. If the evidence seemed to be balanced pro and con, then they had to rule for the defendant, who did not need to prove his innocence nor find the culprit in order to be acquitted. If the evidence indicated that he might not have been at the scene because there was insufficient time to accomplish the deed that had to be accepted.

These considerations have to be made manifest, because during the events that followed the verdict it appeared that the public, educated by the media, did not seem to be aware of these instructions and this trust. Those who indicated they had some awareness of these standards for legal protection for any accused person, often campaigned, after the verdict, for changing the rules, and if that meant adjusting our constitutional rights, they said, it should be done. Such sentiments were inspired, were such inspiration needed, by the televised press conference shortly after the verdict. D.A. Gil Garcetti and his lead prosecutors Marcia Clark and Christopher Darden proclaimed the verdict to be a travesty of justice, adding slurs about those who rendered it. Such questionable behavior by public officers indicated ignorance of the judicial process, for the legally designated "trier of fact" is the jury and we accept jury decisions, looking forward to an appeal if that is desired by the loser.

That was only the beginning of the considerable abuse sustained by jury members, who were accused of being ignorant, of being all African-American (three members were not) and of not understanding the evidence (said by almost everyone who did not follow the trial). Chief Willie Williams used this occasion of resentment against the verdict to state that this verdict proved that the LAPD was slandered in the trial. Many saw the upcoming civil trial as rectification. A different jury, many of whom were from Nicole's neighborhood (where many residents had been quoted in the press as stating that the guilty person had to be Simpson, for if such an act were random, it would deflate property values in the neighborhood). The choice of the Judge (Fujisaki), who retired as soon as the trial was completed, was not left to chance; he did not admit any evidence about the LAPD Laboratory, about LAPD behavior, or anything else that might have been exculpatory for the defendant. Detective Fuhrman was not called to testify, for example. One interesting difference, to this observer, between Judge Fujisaki and Judge Ito: for Fujisaki, not enough trial was too much, and for Ito, too much trial was not enough. News writers, columnists and others cheered the $33 million that Simpson was ordered to pay the families of the victims, but the euphoria quickly subsided when it became clear in the court that no one would ever realize any of this payment. Simpson's financial resources were exhausted paying for defense attorneys and other expenses for both trials. (It is difficult, if not impossible to find a person who does not believe Simpson "bought" this verdict by being able to engage "smart lawyers". Few know that the criminal trial of People v. Simpson cost "the people" (between $8 and $10 million), or that even the lowest estimate was at least twice the amount that Simpson had spent.

Another theme that did not play well with the jury concerned the motives attributed to Simpson during both trials. This centered around themes of jealousy (on Simpson's part) and fear (Nicole's emotion); the former- possessive dark evil and the latter- white innocence defiled at seventeen (a young age), snatched from home and family to be his mistress. However, those who knew both of them well give us a different picture. This had been a very complex relationship, and it is known that after one night with Simpson, Nicole had her family's blessing to "move in" and live with this married man (33). All we know about their relationship were episodes reported as factual by those who had been close to Nicole. Yet, it was Darden's role to put forth this plot and character scene throughout the trial. Jury members paid little attention to it, but the media and the public saw it as central to the denouement.

The bitter outcome of this case for most of our citizens fueled admittance of a variance of the hearsay law which was labeled "letting victims speak from the grave". The variance was passed between the two Simpson trials so that is could have its debut in the forthcoming civil trial. No one appeared to be perturbed by the fact that dead witnesses cannot be cross-examined, and among our citizens (depending on which poll you read), 70% to 80% of white citizens, and 20% to 30% of black citizens felt vindicated and purified by the verdict of liability (in the civil trial) for Simpson, figures which approximate the percentages of those who felt betrayed and saddened by the "not guilty" verdict in the criminal trial.

REVOLUTIONARY IDEAS AND RECIDIVISM

As mentioned, in the aftermath of the criminal trial, many expressed the heartfelt need to have our basic laws changed to reflect what were perceived to be serious shortcomings of our legal system. Among these proposed changes: relinquishing the need for twelve jurors agreeing on the verdict in criminal jury trials, dropping the "reasonable doubt" criterion for the jury to consider, getting rid of the Fourth amendment exclusionary rule (evidence collected without a warrant is inadmissible in court), allowing hearsay evidence in a trial in addition to allowing evidence gathered without a warrant.

Indeed the trial of the People v. O. J. Simpson was a cultural phenomenon, with a complexity of structure and interactions that defied total comprehension by anyone not dedicated to hearing, absorbing, and analyzing all of it; it also demonstrated a great need for an understanding of why our legal system should not be tampered with in the heat of an unpopular verdict. Gerald Uelmen, constitutional expert and scholar, noted that such clarion calls for reform in our justice system are not new in America, giving an example from over a century ago, in which similar reforms were proposed (34). He does not elaborate on his statement that "perhaps the case of the People v. O. J. Simpson is best understood as a cultural phenomenon rather than a lesson plan" (35). We find this an insightful statement, and are particularly pleased that it is proffered, not by an anthropologist, but by an esteemed constitutional expert preparing to write a book for law students on lessons from the trial. His statement offers a good beginning for a cultural approach to this case, as well as to topics in this book that precede it, in our assessment of two systems of logic founded on different premises. For one cannot use the particular language of a subject to try to understand its relevance to other ways of knowing. Next, we will consider two systems of logic in a more general way, exploring the fantasy that they exist divorced from the real world of problem solving.

NOTES

1. *People of the State of California v. Orenthal James Simpson* BAO97211 WL 672670 (Cal. Super. Ct. LA County Sept. 26, 1995) (Sept. 25 end of trial, verdict read Oct. 13, 1995). See also People v. Simpson, no. BAO97211, 1995 WL 394321 at #13 (Cal. Super. Ct. L. A. county, June 23, 1995)
2. Freed, Donald and Raymond P. Briggs. *Killing Time: the first full investigation into the unsolved murders of Nicole Brown Simpson and Ronald Goldman.* New York: Macmillan. 1996. p. 44, 47-48.
3. Attorneys for the prosecution: Gil Garcetti, D.A., Marcia Clark, Christopher Darden, Cheri Lewis, William Hodgman, Lisa Kahn. Attorneys for the defense: Johnnie Cochran, Jr., F, Lee Bailey, Alan Dershowitz, Carl Douglas, Peter Neufeld, Barry Scheck, Robert Shapiro, Gerald Uelmen.
4. Uelmen, Gerald F. Lessons from the Trial: the people v. O.J. Simpson. Kansas City: Andrews and McMeel. 1996. p. 116
5. See Freed and Briggs, p.74
6. Thompson, William C. Proving the Case: the science of DNA: DNA Evidence in the O.J. Simpson trial. 67: *Colo L Review* 827-857 (1996) p. 832
7. Schiller, Lawrence and James Willworth. *American Tragedy; the uncensored story of the Simpson defense.* New York: Random House. 1996 p. 460
8. See Schiller and Willworth p. 447
9. See Schiller and Willworth pp. 446-447
10. See Thompson p. 841
11. See Schiller and Willworth p. 462
12. See Uelmen p. 115
13. See Freed and Briggs p. 273
14. See Freed and Briggs p. 65
15. The preliminary hearing began on June 29, 1994 and ended July 22, 1994
16. Cooley, Armanda; Carrie Bess, and Marsha Rubin-Jackson. *Madam Foreman.* Beverly Hills, California: Dove Books 1995 p. 123
17. Freed and Briggs p. 234
18. See Schiller and Willworth p. 214
19. See Thompson p. 838
20. See Thompson p. 856
21. See Schiller and Willworth p. 485
22. See Romanucci-Ross 1986 in General Bibliography
23. Koehler, Jonathan J. "Proving the Case: the Science of DNA: on conveying the probative value of DNA evidence: frequencies, likelihood ratios and error rates". 67: *Colo L Rev.* 859-886 (1996) pp. 861-862. See also Koehler, Jonathan J. "Error and exaggeration on the presentation of DNA evidence at trial. *Jurimetrics.* J 21 (1993).
24. See Cooley et al.
25. See Singular, Steven. *Legacy of Deception: An investigation of Mark Fuhrman and Racism in the L.A.P.D.* Beverly Hills: Dove Books 1996.
26. Resnick, Faye. *Nicole Brown Simpson: the private diary of a life interrupted.* See also "All American Drugs" pp. 143-169 in Freed and Briggs.
27. See Freed and Briggs pp. 154-157.
28. Darden, Christopher, with Jess Walter. *In Contempt.* New York: Harper-Collins/ Regan Books 1996 p. 33.
29. See Freed and Briggs pp. 275-294 for the report of Dr. Irvin L. Golden, Deputy Medical Examiner.
30. Cooley et al. pp. 4-29.
31. See Freed and Briggs pp. 143-169 " All American Drugs".
32. Dershowitz, Alan. *Reasonable Doubts: the O.J. Simpson case and the criminal justice system.* New York: Simon and Schuster 1996 pp. 21-22.
33. Bosco, Joseph. *A Problem of Evidence: how the prosecution freed O. J. Simpson.* New York: William Morrow 1996 pp. 155-161.
34. See Uelmen p. 38.
35. See Uelmen p. 39.

CHAPTER 9

LOGICS OF DISCOVERY, CHANCE, AND SCIENTIFIC EVIDENCE IN THE COURT

THE NEED TO BELIEVE

In People v. Simpson, understanding the scientific evidence, as well as the legal constraints on presentation of evidence imposed by the court, revealed much about legal and scientific reasoning. Equally compelling was the display of these two knowledge systems as they were creatively reconfigured by adversaries in the trial. As trial events unfolded, the implications of this reconfiguration ran deep and wide throughout the cultural fabric of the country, promoted by the continuous dissemination of information from the trial through television, talk shows, radio, newspapers and tabloids, so that the event came to be considered unique among the previous thirty-two "trials of the century" (1).

Delivered as factual, yet interpretable through cultural themes and racial attitudes, former relationships between the two victims and their respective family members were frequently discussed. Noted and commented upon occasionally were family members of the accused, but the media, without exception, directed its sympathetic focus on the kin of the victims. Families of the victims were idealized; and sisters and parents were afforded complete credibility as they described historical vignettes. However, some writers who were "guests" (given seats by Judge Ito), decided it would be interesting to research selected relationships of one of the families and published different views (2). Themes in "family myths" loomed large as family members were transformed into personalities to be followed in the weekly gossip publications, and, in the same publications, one could find continuing revelations about major players in the trial, particularly the lead attorneys for both sides. Some scientific experts on the defense team had long been declared "stars" in their own right from previous successes in famous trials. This cast, along with several colorful witnesses, lent an aura of glamour to the proceedings. On the periphery of this "center stage", there soon appeared myriad special interest professionals including commentators, columnists, law school deans, professors, psychologists and others, to dissect presentations by experts actually involved in the arena of the court. Such commentaries were incessant, enlightening or obfuscating, depending on the political and cultural agendas of the commentator and the listener. Organizations (some already established and others arising *de novo*) energized by the trial's impact on the public, began fund raising activities. Such organizations included, but were not limited to, those dedicated to social action on such phenomena as spousal abuse, violence inflicted on women by sports figures, parental rights (i.e. denial of such rights to any parent on trial for murder, regardless of the jury verdict), recalling judges who made unpopular decisions (such as awarding child custody to a parent that the group found objectionable, regardless of legal mandates), and lowering the standard of proof in criminal cases. Remarkably, polls demonstrated time after time that most Americans did not follow the trial proceedings; yet this uninformed group, the non-engaged who avoided "direct" information sources such as "court TV", had the strongest, most negative and inalterable opinions about the jury verdict.

Describing what actually transpired in a court of law regarding the evidence, and in particular the scientific evidence, is not the same as holding a brief for Simpson, but invariably many considered this to be the case. This segment of the public remained convinced that the prosecution had a "mountain of evidence", much of it unassailable DNA evidence. They also tended to explain the verdict by the "fact" that the rich buy justice through lawyers who "stage manage" legal trickery and by so-called experts who can conjure up a "win" regardless of the evidence.

Considerations of the forensic evidence dominated the trial overtly, but ultimately impassioned interpretations of human behavior, often scripted and presented by the prosecution attorneys, seemed to "score" with press and public and to challenge the defense mightily. Nevertheless, convinced that their *forte* would continue to be a frontal attack on the scientific foundations of the prosecution's "mountain of evidence", defense lawyers did not dwell on the behavioral aspects of act and intent. The "human behavior" explanatory model proffered by the prosecution was focused either on Simpson's possessiveness and jealousy, or on his dream revelation to another individual, or on a "dark mood" perceived by several members in the audience at his daughter's dance recital. This assertion by a number of witnesses was quickly dispelled by several videos showing a jovial, happy Simpson on friendly affectionate terms with his in-laws. Such evidence was rebutted in terms of "possession" by a sinister spirit (of the jovial person the public had long known), with an indication that the possessed Simpson was the "real" Simpson. However, forays into manner and motive did not appear to have an appreciable effect on the jury, according to those jurors who subsequently wrote about it, but apparently had a major effect on many in the public (often referred to as "the court of public opinion") who seemed to find in these unprovable assertions the determinants of probable cause for the murders.

The anthropologist (Lola Romanucci-Ross) found herself repeatedly drawn into post-trial discussions on no fewer than seventy occasions in various parts of the country and in several countries in Europe and Asia. Such discussions arose spontaneously during breaks at scientific meetings with other academics, or on social occasions, (both in and out of such sessions), and were always initiated by others. (Notes were taken and kept, as habits of anthropologists do not allow holidays from "the field"). In foreign countries, the questioning appeared to be informed, but was usually followed by the apology, "of course, I know only what the newspapers or radio or TV commentators tell me", sometimes followed by, "I did not methodically explore opportunities to learn more or to check the facts". They thought that since Simpson was presumed guilty by all the journalistic and television sources from America, it must be true. Persons in foreign countries were very surprised to hear replies to their questions, and to learn that some events that took place in the courtroom did not square at all with what they had read and heard in the media concerning those events. But persons in the United States with admittedly similar sources of information had a markedly different reaction to disparities between what the court determined as facts, and what the media reported as factual. Here, in the responses of discussants, one found annoyance, disbelief, anger and, at the very least, discomfort. The disappointing outcome of the trial was always explained in terms of trickery by the defense team, by the intellectual shortcomings of the jury, and/or the weakness of the Judge, who nurtured things along so badly that these events were allowed to determine the verdict of acquittal. Indeed, it seemed there were two different trials (3), one in "the court of public opinion", scripted by the media, and one in a Los Angeles court room, presided over by Judge Lance Ito.

The "trial in the court of public opinion" began quite some time before the trial in the LA courtroom. The Police Department (LAPD) began its campaign for conviction by releasing the famous "911" tape, which recorded a phone call from Nicole, made a number of years before the murders, requesting that Simpson be removed from her house. Few in the public could remember, or had even noticed what was actually said on the tape, nor did they know or note that she then told the police to forget about it, for he had left quietly. As the trial progressed, there were a number of press releases from the same sources in the LAPD, containing references to blood samples, along with results of tests that had not yet taken place. Yet there were themes running through the trial about police who never lie, are always fair and objective, with no other goal but pursuit of the truth. But not so in the characterizations of the jurors, who could be prejudiced, particularly if they are Black.

THE ARENA, AND "CHANCE"

> "Honda, however, knew something about what went on inside a judge. And how intense were a judge's inner struggles! Emotions of sentiment, desire, personal concern, ambition, shame, fanaticism, and all sorts of other flotsam—the fragments of planks, the waste paper, the oil slick, the orange peels, the fish, the seaweed fillers, the sea of human nature that was ever pushing against the lone seawall that kept it in check: legal justice. Such was the struggle" (Mishima)

No arena of any struggle is ever free of the element of chance. In this trial, as in any other, the power of the Judge was manifest, for a Judge decides, as is his prerogative, whether something is material or collateral, probative or prejudicial (whether the inflammatory nature of the evidence exceeds its probative value), whether it is relevant or "beyond the scope", whether it is evidentiary or hearsay. Now "the court" also decides what is and what is not "scientific evidence". Therefore, the Judge controls, defines, and limits the logical system within which the debate may be heard and considered by the jury, as well as other observers of the trial. (This is the sort of power that the rest of us can only dream about!).

The judge eventually chosen to preside over a trial is a factor that both sides have to accept. Like most judges, Judge Ito had been a prosecuting attorney and judges are either appointed or elected. In general, defense lawyers are not chosen for this position. Superior court judges in California are appointed by the Governor for a six-year term, after which they must stand for re-election countywide and even without the pressure of re-election, a judge ("the court") is part of the state's structure of justice. The District Attorney, the Police Department, the Department of Justice and the FBI, etc. all have the resources of the state at their disposal. To be assigned to preside over a trial may or may not be a matter of chance for the judge, but the eventual choice is a "chance" element for both sides, or at least for one side. Judge Ito, although he denied over two-thirds of motions put forth by the defense, was the best choice the defense could have hoped for; he was learned, very intelligent, highly competent, and sincerely dedicated to making the correct "calls", so that the verdict would not be overturned by the appeals court. Had the verdict been the one devoutly wished for by many, then the prosecution, the media pundits, and others would have recognized Judge Ito's hard work and critical attention to many serious problems he had to solve during the trial.

Judges are at times targets of both journalistic and public criticism, but the police are generally beneficiaries of public trust. Detective Mark Fuhrman, while officially "off the case", in unsupervised moments managed to find evidence of considerable import, i.e., the bloody glove in Simpson's backyard. He asserted that the glove was found during a warrantless search after he jumped over the fence at the Simpson estate, telling his police colleagues that the grounds ought to be checked for other possible victims, or perhaps to prevent someone from becoming the next victim. Because of the impact of the glove on the case, the defense had a challenge: they did not believe that the glove was found, but rather that it was planted earlier by Fuhrman and then "found". But to assert this would have accomplished nothing, so they hoped to uncover evidence about Fuhrman to make this possibility credible. They learned that he had been interviewed over a period of time, in the past, by an aspirant writer who was planning to write a screenplay about women in the LAPD. After the expensive adventure of a legal foray to gain access to the tapes (Johnnie Cochran had to take the request to court in North Carolina, to appeal the decision, and then win the appeal), the defense did gain access to the tapes and asked to have some material from them admitted as evidence. There were many hours of Fuhrman on these tapes giving accounts of police behavior in the Ramparts Division of the LAPD. He spoke of officers planting evidence, telling lies on the stand if necessary to win a conviction and to doing "what it takes" to incarcerate a suspect whom they felt or "knew" was guilty. Fuhrman asserted on the tapes that this behavior was quite common, at least in the Ramparts Division. Very little of this material was admitted into the trial by Judge Ito: he allowed only two innocuous comments by Fuhrman that gave the lie to his assertion on the stand that he had never used the word "nigger" (referring to Blacks) in the previous ten years.

The myth that law enforcement officers were incapable of deception was discredited, but this was not easily accepted by the public, and for most it appeared effortless to explain this away by wishing to believe that Fuhrman fantasized it all so that the "lady writer" would have material for her screen play. Later, this explanation was destroyed by newspaper accounts of actual occurrences, validating Fuhrman's descriptions. In fact, in the years that followed the Simpson trial, the Los Angeles Times published a number of articles on such behavior by officers in the Ramparts Division; episodes of evidence planting, framing of suspects by officers, and cover-ups of unjustified shootings, all of which were established as factual by evidence presented at trials (4).

MIND AND EXPERIENCE

As we indicated earlier, ideology shapes the way we, or members of any other culture, view the world and structure the categorizing and understanding of all experience. It provides the grid for how we experience and interpret not only what we see and hear, but also what is scotomized, that is "blacked out". If events find neither a niche for interpretation, nor a "fit" into the schema or schemata of what we already know, information does not "make sense". After the Simpson verdict, all polls showed that anywhere from 67% to 80% (depending on the poll) of White Americans believed that Simpson murdered Nicole and Ron Goldman. Among Black Americans, the reverse was true, with the substantial numbers in the majority believing him not guilty of the crime. According to various polls, there was general agreement that only about 4% of the public watched most or all of the trial.

Certain events preceding, accompanying, occupying, or even foundering in the wake of the trial abounded in myth, and we use the term in the manner it was understood and used by the Ancient Greeks. In this sense, a myth is a truth more real than empirical; also, it is accepted as true even if it never happened, and even if we do not understand why we want to believe in it. It is true because believing in it determines human behavior, which then makes the myth "real" in the sense that we want to have truth "realized". In terms of a heavy investment in myth, the Simpson matter (as Judge Ito referred to each day's session) was a drama of mythic proportions, a theme park about money, power, enchantment, celebrity, and, at least in some minds (fueled by a Chris Darden prosecution "theme") about love found, love lost, and jealousy. To many Americans, the famous and beloved running back O.J. Simpson had been larger than life, more gifted than any man is entitled to be, and endlessly successful in many endeavors beyond football. In following his biography we now had arrived at the Romance of the Fallen Hero on trial for his life or his freedom, with his support base eroding. Deep in the human group consciousness, were we reliving a late twentieth century version of "The Golden Bough"?

Sir James Frazer wrote his twelve volume anthropological classic of that title (5) on a research mission: he wanted to study the longing of a cultural group for a hero, who then had to be destroyed. The hero was, among all others, to be the most handsome, most intelligent, most agile and battle-worthy guardian of the mistletoe on the top branch of the sacred oak tree, ever aware that someone would one day slay him, climb the tree, take possession of the mistletoe, and replace him as hero. The new hero and guardian of the sacred oak would now be celebrated in the hearts of every member of the group and be the new recipient of their generosity for all his needs and desires. Coronation of the new hero was an occasion filled with pride and admiration; it provided a solemn occasion to demonstrate group solidarity, and to assure fertility of plants and people for continuity of group existence. But this ceremonialized renewal had to occur through the demise and death of the former hero, a destiny he knew would be his at the first sign of his weakness. Frazer's research began with the sacred oak tree at Lake Nemi in Italy and followed analogous instances of similar examples of social renewal through loss of strength of the hero, and continued with "regicide" throughout Europe in the following centuries.

How did people in such societies and cultures progress from believers in magic to believers in science? Frazer offers his analysis: the fundamental presupposition of magic is actually identical to that of science: "faith implicit, but real and firm in the order and uniformity of nature" (6).

THE IDEOLOGY OF SCIENTISM AND DNA

Not many, if any, of us considers himself or herself ideologically driven. However, ideology is a condition of learning and knowledge that does not contain isolated compartments labeled "religious" or "political" or "scientific". It includes attitudes and inconsistencies which believers learn to live with as they emphasize differences from other belief systems (7). Ideology is acquired during the process of becoming an adult member of one's culture through, for example, socializing institutions such as families, schools, clubs, churches, sports programs and through advertising and other narratives that become part of our experience. It determines and shapes what we classify and how we interpret what we classify. Because of this, every "world view" within any culture has its own idea of what a fact is, and a precise theory on the nature of that fact. Consequently, even if a fact is free of all doubt-inducing theory, it is never free of some modest insight or formulation to anchor its precision and accuracy. A fact is not perceived without a cognitive unifying sensation, perception, imagination, inquiry, insight, formulation, and reflection. If rational consciousness cannot criticize itself because the investigator is involved, engaged, and committed, then where, exactly, do we find the disjunction between rational consciousness and irrational man? Furthermore, significance accrues from insight not from data, and "relevance" is not a property of things or acts; it is to be found in the mind of the beholder. Individual belief systems have an internal logic of their own, and they are not fragmented and labeled for different circumstances. As has been noted, believing in the infallibility of the Pope is at once a statement about morality, about politics, about biology, about attitudes toward science (8).

If science is polysemic, as we earlier asserted, then the word "science" has a multiplicity of meanings which have been building upon earlier concepts, slowly creating a new edifice of meanings, shaped to accommodate the needs of groups that want the meaning of science to fit their views and purposes. With such definitions of science, we are left to debate the lingering inferential interpretations of "science" that overflow the boundaries that were meant to contain them.

Stimulated by hearings set by the House Committee on Science and Technology on March 31 and of April 1 of 1981, Broad and Wade describe the "conventional ideology of science", emphasizing that accepting it as factual makes it impossible to explain fraud; this disconnect is the subject of their inquiry (9). Albert Gore, Jr., Chair of the Committee, wanted the hearings to be an investigation of fraud in scientific research; to this end he had called several respected senior scientists, such as Philip Handler, then president of the National Academy of Science and a leading spokesman for the scientific community. Handler brushed off allegations of fraud as merely exaggerations by the press, maintaining that fraud hardly ever occurs, and when it does, it is immediately discovered, because, he lectured the congressmen, "that is how science works". Many congressmen expressed annoyance at the *hauteur*, and the condescending attitude of the scientists, so certain of the practice of the ideals of their trade, and piqued at the idea that the way it operates would be called into question, or that scientific results might be considered untruthful. Just a few weeks after these hearings, the "fraud" case of John Darsee, involved in laboratory cardiology at the Harvard University School of Medicine came to light (10), providing an occasion not only to observe how fraud can have quite a lengthy career in august surroundings, but also how research mentors and institutions can procrastinate and obfuscate in responding to allegations, even after such allegations have been proven correct. In the face of undeniable proof that in the operational mode science does not necessarily work as described and advertised, many spokesmen, mentors and sponsoring institutions nevertheless continue to hold that there is no need to put safeguards in place. As in the cargo cults of New Guinea (11), the ideology had become the reality for them. When the "cargo" does not arrive, as the true believers are sure that it will, it is never the belief system that is examined for flaws. Rather, the failure is always explained by some personal shortcoming of a participant, or an event, that will never be repeated.

This particular ideology about science has been nurtured and promulgated over many decades by historians, philosophers, and sociologists who have given serious thought to the workings of science and to describing their perception of it. The following are some of the characteristics of the ideology of the discipline and practice of science with which many of us have been imbued from high school through graduate studies: "science is self-correcting"; the goal of scientists is to "pursue the truth wherever it may lead"; and scientific knowledge is always validated by these two pillars of the scientific ethic. Broad and Wade give numerous detailed examples of "fraud in halls of science" (both in our times and in times long past) to illustrate their contention that to study fraud in science is to understand how science "really works": it is empirical, pragmatic, a trial-and-error procedure in a field which is highly competitive, thereby driving its practitioners from one technique to another to achieve success, and often resorting to rhetoric for persuasion. Above all, there is no one scientific method; "science is a social process... and a cultural form that allows the fullest opportunity for the expression of the human propensity for rational thought" (12), a statement with which any anthropologist should be in complete agreement.

"DNA FINGERPRINTING": END LINKAGES

In the early days of DNA technology and exploration of its uses for precision in identification, it seemed logical for many scientists to refer to such uses as "DNA fingerprinting". To equate this new DNA analysis technique with fingerprinting, suggested that it had indisputable accuracy and exclusivity, i.e. identification of persons with no possibility for error. One point of similarity in both methods was that both fingerprint evidence and DNA evidence were dependent for accurate identification on unique characteristics of an individual, leading to the declaration of a "match".

We find the phenomenon of "end linkage" in such early attempts to label the innovative process (i.e. "DNA fingerprinting") in order to convey to the public the scientific certitude with which a "match" might be declared. In creating an "end linkage", one uses, in this case, a metaphor to compel a recent methodology to assume the attributes (say, of certainty in identification) of an earlier, established one. Fingerprints at a crime scene had long enjoyed that distinction, rightly or not. Here we saw the use of a literary mode to equate and therefore validate the new scientific endeavor. Linking concepts from these two systems of identification (both lauded in courts), reflects the power of such a literary license to buttress a system by endowing it with a logical connection.

Receptors for this process in this case were strong; the public has a considerable investment in belief in the power of science in the search for "truth", and perhaps there is even a more understandable investment in the belief that the courts are systematized guaranteurs of fairness and justice. Both systems, law and science, then, are "rational". These beliefs are salient in Western culture and they manage to co-exist with faith in a teleology of Divine Intent.

Fingerprint evidence was thought to be thoroughly reliable; yet there were those who were skeptical. Accuracy is critical in forensic cases, but usually police "dust" only fragments for evidence, and often these are contaminated. Even though the FBI has a database of 44,000,000 sets of prints, the information to be gleaned from forensic samples consists only of an approximation of the best possible "match"; of course matches are declared by people hired to declare them, after visual analysis yields a subjective judgment. It may take hours (sometimes days) to declare a match; moreover, there has never been a study of precisely what is required to declare a match (13).

There were, and still are, some caveats that do not seem to preoccupy many practitioners of either fingerprinting or DNA analyses. We refer specifically here to the forensic situation in which such matches are made and the forensic situation, of course, usually differs greatly from that in contract laboratories. We also add that in both types of analysis, attention should be focused on the size and diversity of the reference database, which, as we indicated earlier (Chapter 8), can be problematic. Thus, a database that is too restricted may suffice for excluding but not for including a suspect.

FINGERPRINTING IN OUR TIMES

On January 3, 2002, a federal judge disallowed the use of a fingerprint expert's opinion in a Philadelphia murder case. There had been previous episodes of legal challenge that argued that "fingerprinting" did not have sufficiently rigorous standards for admission as scientific evidence in a case. Some noted that during 100 years of use in the courts there had been no rigorous research on fingerprinting technique, no determination as to whether the method was valid or even testable, no commentaries on acceptable error rates and no published peer-reviewed article on any of these aspects (14). The above-mentioned Judge (Pollack) entertained such thoughts, and then incorporated them into an official court commentary. In his January 2002 opinion, he wrote that the Daubert standard which asked "is it testable?" could not be the standard for presenting fingerprint evidence, and declared that fingerprint evidence could be presented as" forensic science". Later, in March 2002, noting that mistakes in identification had been discovered (15), he decided that fingerprint analysis evidence was a specialty that did not reach the level of science. However, he added that a technical or specialized expertise need not do so.

THE IDEOLOGY OF SCIENCE, AND FINGERPRINT EVIDENCE

Even today, fingerprint evidence is considered important in solving crimes, and determinant if one finds a "match". How does one find a match? One common method is to compare the width of the ridges, noting direction of flow, and finding points of similarity in bifurcations. Yet no experiment has ever been done on the reliability of such a "match". The tests offered by the IAI (International Association of Identification) which only professionals take, are repeatedly flunked by 50% of those who take it. (Passing it is not a requirement for certification!) (16). Crime labs and judges do not ever ask if one is certified as a fingerprint expert before admitting testimony.

As Faigman points out, there is a vanishingly small likelihood that a given set of ridge characteristics (say 8, 10, or 12) on a given fingerprint will be found in a random sample of the population. But the statistical hypothesis depends on what proportion of the population (or relevant subpopulation) has these particular ridge characteristics (19). Also, various countries have different standards for the number of points of similarity, so that a "match" can be declared in Brazil with 30, Sweden with 7, Italy with 17, Australia with 7.

Injecting attributes from the never- before- questioned fingerprint evidence onto the new technology of DNA analysis was a literary link, rather than an application of a scientific principle. Yet fingerprint evidence, like DNA, could be precise if adequate laboratory conditions obtain and the work is done by well-trained investigators and analysts, leaving few, if any, aspects open to interpretation.

ADMISSIBILITY

The Frye rule (18) determines admissibility for generally accepted scientific evidence, recognizing that a scientific principle at some point "graduates" from experimental to demonstrable, thereby qualifying as "evidentiary". The principle from which the deduction is made must be sufficiently established to have gained general acceptance within the relevant field. Thus, Frye did not allow for the admission of expert opinion in novel scientific areas when the users of a new technique (crime laboratories for the most part) had accepted it regardless of validation. Also, it was sometimes difficult to identify the proper membership field for such novelties (e.g. what to do with "voice-printed" techniques developed at Bell Laboratories in New Jersey).

Among its considerations, Daubert (19) questioned whether the Federal Rules of Evidence of 1975 (especially rule 702) supported the Frye Rule. As a result, three major changes emerged. First, a change in process establishing that trial judges would be "gatekeepers", charged with deciding, prior to trial, the admissibility of evidence which one side had challenged as unreliable. Secondly, it put forth a categorical statement that passage of the Federal Rules of Evidence had done away with the Frye test of "general acceptance in the field to which it belongs", as the sole basis for admissibility. Thirdly, it indicated that the standard to be used henceforth in the federal courts for general acceptance would be proof of reliability (a criterion for admissibility issue under Federal Rules of Evidence 702).

The court also offered guidance as to how proof of reliability was to be established: 1) Is the theory of the technique testable and has it been tested? 2) Has the theory or technique been subjected to and did it survive peer review and publication? 3) Is there a known or potential error rate? 4) Is there "general acceptance"?

Based on subsequent decisions, the states—some of which have rules of evidence patterned upon the federal rules—were free to follow the Daubert approach or to reject it. A number of states did adopt the Daubert factors in their approach to decide admissibility of expert opinion testimony. Surprisingly, however, the majority of states chose to retain the Frye test as their guiding principle for the admission of expert testimony. In almost all of the state jurisdictions that elected to retain Frye, recent decisions have also explored whether the Daubert factors were satisfied as well, even while insisting Frye remains the law in the state.

In a 7-2 decision Supreme Court, Chief Justice William Rehnquist did not approve of the new guidelines for judges, saying that they would turn judges into amateur scientists. Although we have seen that Daubert and subsequent "refinements" may have kept some substandard science out of the courtroom, many observers have noted that judicial discretion (as Daubertian "gatekeepers") have managed to keep out "good science" as well (20).

SOME SURPRISING AFTERTHOUGHTS

We noted some examples of the exercise of discretion in the previous chapter (relevant to statistical frequencies). Judges tend to dismiss testimony from physicians as anecdotal, and in such instances often set standards for scientific evidence other than those which doctors or medical scientists use in their daily research. Also, judges do not easily tolerate scientific disagreement, and they tend to dote on epidemiological evidence. However, there are many problems with that kind of scientific evidence (21) which most of them do not appear to recognize. Often it is the type of medical science that such judges find does not pass the Daubert Test that has made great strides in medicine. Also, disagreement among medical scientists is not uncommon, and, such disagreements often lead to much progress. The jury remains the trier of fact, according to the law, and it can be held that they should be able to consider such testimony.

The heavy reliance by judges on epidemiology is rather odd, considering that epidemiology is not really a science, but rather a "technique" for probing possible cause and effect relations through the study of populations. To us perhaps, the most fundamental flaw in the epidemiological endeavor is what is left out of the original study design: epidemiologists measure what they can measure, count what they can count. This often leaves out what may later turn out to have been the most important variables. To correct for such possibilities, statisticians apply their own methods, and rules for manipulating the data. It cannot be easy for a judge to follow all of this, as we exemplified in the Simpson case in the previous chapter.

In Chapters 5 through 8, we considered three case studies to illustrate the range of analytical difficulties inherent in the shaping of scientific information by agents who have social goals for policy decision-making. The example of the sexual predator offers an illustration of one end of the spectrum, where strictly biochemical and physiological indicators play a relatively small role in framing scientific information to meet the requirements of a social policy decision. There are a number of cultural and biological factors which define values such as normality, abnormality, abuse, addiction and sexual preference, all of which are open to multiple interpretations, and are heavily influenced by cultural differences. In the case of brain imaging (Chapter 6), we find much "hard data", and learn about the major role of interpretation of those data. Imaging technologies (e.g. Positron Emission Tomography) reveal alterations in brain function at the time that they are done, and such findings are not the source of major disagreement. The meaningful question, however, is: what do the data mean when used for interpreting human behavior, which is importantly influenced by many factors that are not biological? In discussing DNA (Chapter 7), we find the analyses are often highly reliable, but nevertheless we note that there are shifting standards in forensic as opposed to research laboratory methods; in the latter there is more than enough interpretation and quite often confusion as to the importance of identified differences. Discussion of the Simpson case (Chapter 8), provided many examples from the trial of the cognitive modeling of two professional systems of discourse in an arena in which these two systems are in confrontation. In considering all four examples, we discussed sets of rules which stem from thought systems that, although they meet in the arena of the court, do not match, do not mesh, do not "make sense" in complementarity, because their origins have different premises and different goals. We also have considered how these differences can be identified and might be resolved.

There is no one "scientific method". Science is about Inquiry, as are many other ways of knowing. As in other fields, trades or professions (22), there are professional organizations, editors of journals, reviewers, committees that recognize meritorious achievements and all of this social and political activity ultimately has a defining influence on what becomes acceptable as science. Is there a logical system peculiar to science, or, as Descartes held, a method? "Method" includes techniques such as hypothesizing, experimenting, measuring, modeling, theorizing, and predicting. Most people would like to believe that every scientist practices these techniques in that order. This work has attempted to address the complexity of that issue, by introducing material from other cultures, of different levels of complexity and sophistication, and finding analogies in our own culture. It has also put the scientific method under close scrutiny in adversarial situations in the courtroom. It is to be hoped that this approach will have a beneficial effect on how scientists, as well as lawyers, judges and the rest of society views the scientific endeavor in this context.

NOTES

1. Gerald F. Uelmen. *Lessons from the Trial: The People v. O.J. Simpson*. Kansas City: Andrews and Momeet, 1996, p.32.
2. Bosco, Joseph. *A Problem of Evidence: How the Prosecution Freed O.J. Simpson.* New York: William Morrow and Co. 1996, pp. 155-161 ("The Family Brown").
3. Dershowitz, Alan M. *Reasonable Doubts*. New York: Simon and Schuster. 1996, pp. 129-132.
4. *L.A. Times,* Mar 1, 2003. B4, cols 1-4. (LA DA Steve Cooley wants to get rid of the Ramparts investigation of 82 Ramparts related cases); *LA Times* Oct 5 2003 (New Chief Bratton wants a Blue Ribbon Committee to investigate the framing of suspects and unjustified planting); *LA Times* April 23, 2003 Sec. B p. B15 cols 1-6 (by Erwin Chemerinsky. "To Prevent a Repeat of Ramparts, Fix More than the LAPD").
5. Sir James Frazer. *The Golden Bough: A Study in Magic and Religion.* 12 Volumes published from 1907 to 1915.
6. Frazer vol. 1 p. 61.
7. Lola Romanucci-Ross, D. Moerman and L. Tancredi. *The Anthropology of Medicine from Culture to Method.* Bergin and Garvey (London and Westport, Conn). 1997 (3rd edition), pp. 357-361.
8. Keach, Milton. *The Open and Closed Mind: investigations into the nature of belief systems and personality systems.* (In collaboration with Richard Bonier and others). New York: Basic Books, 1960.
9. William Broad and Nicholas Wade. *Betrayers of the Truth; Fraud and Deceit in the Halls of Science.* New York: Simon Schuster, 1982.
10. Broad and Wade. Op. Cit. pp. 13-15.
11. Schwartz, T., Romanucci-Ross, L. and M. Mead. References op. cit. in chapter one notes.
12. Romanucci-Ross and Moerman op. cit. chapter one notes p. 351-2.
13. "Fingerprints" CBS "Sixty Minutes" 01/05/03.
14. David L. Faigman. "Is Science different for lawyers?" *Science*. 27:19:339-340, July 2002.
15. *United States v. Plaza*, 179 F. Supp. 2D 492 (Ed. Pa 7 January 2002).
    United States v. Plaza, 188 F. Supp. 2D 549 (Ed. Pa 13 March 2002 Plaza 11).
16. "Fingerprints" op. cit.
17. Faigman, op. cit. p.340.
18. *Frye v. United States* 293 F. 103 (D.C. Cir. 1923).
19. *Daubert v. Merrell Dow Pharmaceuticals, Inc.*, 509 U.S. 579 (1993).
20. Sharon Begley. "Ban on 'junk science' also keeps jurors from sound evidence. *Wall Street Journal*, June 27, 2003. Section B1 Page 1, Col. 1.
21. Begley, op. cit.
22. For example, for the teaching profession, see Ivan Brady and Alok Kumar "Some thoughts on Sharing Science" in *Science Education*. June 2000. pp. 507-523.

APPENDIX

FRYE v. UNITED STATES

No. 3968[1]

COURT OF APPEALS OF DISTRICT OF COLUMBIA
293 F. 1013; 1923 U.S. APP LEXIS 1712; 54 APP. D.C. 46; 34 A.L.R. 145
SUBMITTED NOVEMBER 7, 1923
DECEMBER 3, 1923, DECIDED

PRIOR HISTORY: [**1]
Appeal from the Supreme Court of the District of Columbia.
OPINIONBY: VAN ORSDEL
OPINION: [*1013] Before SMYTH, Chief Justice, VAN ORSDEL, Associate Justice, and MARTIN, Presiding Judge of the United States Court of Customs Appeals.

VAN ORSDEL, Associate Justice, Appellant, defendant below, was convicted of the crime of murder in the second degree, and from the judgment prosecutes this appeal.

A single assignment of error is presented for our consideration. In the course of the trial counsel for defendant offered an expert witness to testify to the result of a deception test made upon defendant. The test is described as the systolic blood pressure deception test. It is asserted that blood pressure is influenced by change in the emotions of the witness, and that the systolic blood pressure rises are brought about by nervous impulses sent to the sympathetic branch of the autonomic nervous system. Scientific experiments, it is claimed, have demonstrated that fear, rage, and pain always produce a rise of systolic blood pressure, and that conscious deception or falsehood, concealment of facts, or guilt of crime, accompanied by fear of detection when the person is under examination, [**2] raises the systolic blood pressure in a curve, which corresponds exactly to the struggle going on in the subject's mind, between fear and attempted control of that fear, as the examination [*1014] touches the vital points in respect of which he is attempting to deceive the examiner.

In other words, the theory seems to be that truth is spontaneous, and comes without conscious effort, while the utterance of a falsehood requires a conscious effort, which is reflected in the blood pressure. The rise thus produced is easily detected and distinguished from the rise produced by mere fear of the examination itself. In the former instance, the pressure rises higher than in the latter, and is more pronounced as the examination proceeds, while in the latter case, if the subject is telling the truth, the pressure registers highest at the beginning of the examination, and gradually diminishes as the examination proceeds.

Prior to the trial defendant was subjected to this deception test, and counsel offered the scientist who conducted the test as an expert to testify to the results obtained. The offer was objected to by counsel for the government, and the court sustained the objection. [**3] Counsel for defendant then offered to have the proffered witness conduct a test in the presence of the jury. This also was denied.

Counsel for defendant, in their able presentation of the novel question involved, correctly state in their brief that no cases directly in point have been found. The broad ground, however, upon which they plant their case, is succinctly stated in their brief as follows:

"The rule is that the opinions of experts or skilled witnesses are admissible in evidence in those cases in which the matter of inquiry is such that inexperienced persons are unlikely to prove capable of forming a correct judgment upon it, for the reason that the subject-matter so far partakes of a science, art, or trade as to require a previous habit or experience or study in it, in order to acquire a knowledge of it. When the question involved does not lie within the range of common experience or common knowledge, but requires special experience or special knowledge, then the opinions of witnesses skilled in that particular science, art, or trade to which the question relates are admissible in evidence".

Numerous cases are cited in support of this rule. Just when a scientific [**4] principle or discovery crosses the line between the experimental and demonstrable stages is difficult to define. Somewhere in this twilight zone the evidential force of the principle must be recognized, and while courts will go a long way in admitting expert testimony deduced from a well-recognized scientific principle or discovery, the thing from which the deduction is made must be sufficiently established to have gained general acceptance in the particular field in which it belongs.

We think the systolic blood pressure deception test has not yet gained such standing and scientific recognition among physiological and psychological authorities as would justify the courts in admitting expert testimony deduced from the discovery, development, and experiments thus far made.

The judgment is affirmed.

WILLIAM DAUBERT, ET UX., ETC., ET AL., PETITIONERS V. MERRELL DOW PHARMACEUTICALS, INC.

NO. 92-102[2]

SUPREME COURT OF THE UNITED STATES

509 U.S. 579; 113 S. CT. 2786; 125 L. ED. 2D 469; 1993 U.S. LEXIS 4408; 61 U.S.L.W. 4805; 27 U.S.P.Q.2D (BNA) 1200; CCH PROD. LIAB. REP. P13,494; 93 CAL. DAILY OP. SERVICE 4825; 93 DAILY JOURNAL DAR 8148; 23 ELR 20979; 7 FLA. L. WEEKLY FED. S 632

MARCH 30, 1993, ARGUED
JUNE 28, 1993, DECIDED

PRIOR HISTORY: ON WRIT OF CERTIORARI TO THE UNITED STATES COURT OF APPEALS FOR THE NINTH CIRCUIT.

DISPOSITION: 951 F.2d 1128, vacated and remanded.

DECISION:

"General acceptance" of principle underlying scientific evidence held not to be necessary precondition to admissibility of such evidence under Federal Rules of Evidence.

SUMMARY:

A minor child and his parents, together with another minor child and his mother, brought suit in a California state court against a drug company which had marketed the prescription drug Bendectin. The plaintiffs alleged that the children's birth defects had been caused by the mothers' ingestion of Bendectin during pregnancy. The suit was removed, on diversity grounds, to the United States District Court for the Southern District of California. The company moved for summary judgment and submitted, in support of the motion, the affidavit of an epidemiologist to the effect that no published epidemiological (human statistical) study had demonstrated a statistically significant association between Bendectin and birth defects. In response, the plaintiffs offered expert opinion testimony based on (1) test-tube and live-animal studies that had allegedly found a link between Bendectin and birth defects; (2) pharmacological studies that allegedly showed similarities between the chemical structure of Bendectin and that of substances known to cause birth defects; and (3) the reanalysis, or recalculation, of previously published epidemiological studies. The District Court, granting summary judgment in favor of the company, expressed the view that (1) scientific evidence is admissible under the Federal Rules of Evidence only if the principle on which such evidence is based is sufficiently established to have general acceptance in the field to which it belongs; (2) epidemiological studies were the most reliable evidence of causation of birth defects; (3) the testimony based on test-tube, live-animal, and pharmacological studies was inadmissible because such testimony was not based on epidemiological evidence; and (4) the testimony based on reanalyses was inadmissible because the reanalyses (a) apparently had never been published or subjected to peer review, and (b) failed to show a statistically significant association between Bendectin and birth defects (727 F Supp 570). The United States Court of Appeals for the Ninth Circuit, affirming on appeal, expressed the view that (1) expert opinion based on a scientific technique is inadmissible if the technique is not generally accepted as reliable in the relevant scientific community; and (2) under the general acceptance standard, the plaintiffs' evidence provided an insufficient foundation to allow admission of expert testimony that Bendectin caused birth defects (951 F2d 1128).

On certiorari, the United States Supreme Court vacated the Court of Appeals' judgment and remanded the case for further proceedings. In an opinion by Blackmun, J., expressing the unanimous view of the court as to holding 1 below, and joined by White, O'Connor, Scalia, Kennedy, Souter, and Thomas, JJ., as to holdings 2 and 3 below, it was held that (1) the "general acceptance" test of Frye v United States (1923) 54 App DC 46, 293 F 1013, 34 ALR 145, was superseded by the Federal Rules of Evidence (FRE), and thus general acceptance is not a necessary precondition to the admissibility of scientific evidence under the FRE, given that (a) nothing in the text of Rule 702 of the FRE, governing expert testimony, establishes general acceptance as an absolute prerequisite to admissibility, and (b) there is no indication that Rule 702 or the FRE as a whole were intended to incorporate a general acceptance standard; (2) under the FRE, a federal trial judge must insure that any and all scientific testimony or evidence is not only relevant but reliable; and (3) in a federal case involving scientific evidence, evidentiary reliability is based on scientific validity.

Rehnquist, Ch. J., joined by Stevens, J., concurring in part and dissenting in part, (1) agreed that (a) the Frye "general acceptance" rule did not survive the enactment of the FRE, and (b) Rule 702 of the FRE confides to the trial judge some gatekeeping responsibility in deciding questions of the admissibility of proffered expert testimony; but (2) expressed the view that the Supreme Court should have left the further development of the area of the law in question to future cases.

LAWYERS' EDITION HEADNOTES:

[***LEdHN1]

EVIDENCE §641

expert scientific testimony -- admissibility -- general acceptance standard --

Headnote:

A standard under which the exclusive test for admitting expert scientific testimony is whether the principle on which such testimony is based has general acceptance in the field to which it belongs is not to be applied in federal trials; the "general acceptance" test of Frye v United States (1923) 54 App DC 46, 293 F 1013, 34 ALR 145, is superseded by the Federal Rules of Evidence (FRE), and thus general acceptance is not a necessary precondition to the admissibility of scientific evidence under the FRE, given that (1) nothing in the text of Rule 702 of the FRE, governing expert testimony, establishes general acceptance as an absolute prerequisite to admissibility; and (2) there is no indication that Rule 702 or the FRE as a whole are intended to incorporate a general acceptance standard, as (a) the drafting history makes no mention of the Frye decision, and (b) a rigid general acceptance requirement would be at odds with the liberal thrust of the FRE and their general approach of relaxing the traditional barriers to opinion testimony.

[***LEdHN2]
COURTS §538.11
construction of rules --
Headnote: [2]
A court properly interprets the legislatively enacted Federal Rules of Evidence as the court would interpret any statute.

[***LEdHN3]
EVIDENCE §641
expert scientific testimony -- reliability --
Headnote: [3A] [3B] [3C]
Under the Federal Rules of Evidence (FRE), a federal trial judge is not disabled from screening purportedly scientific evidence; rather, the trial judge must insure that any and all scientific testimony or evidence admitted is not only relevant but reliable; the primary locus of this obligation is Rule 702 of the FRE, which governs expert testimony as to scientific knowledge; for purposes of Rule 702, "scientific" implies a grounding in the methods and procedure of science, and "knowledge" connotes more than subjective belief or unsupported speculation; although it would be unreasonable to conclude that the subject of scientific testimony must be known to a certainty, Rule 702 requires that proposed scientific testimony be supported by appropriate validation--that is, good grounds--based on what is known; Rule 702's requirement that an expert's scientific testimony pertain to "scientific knowledge" establishes a standard of evidentiary reliability, that is, trustworthiness; in a federal case involving scientific evidence, evidentiary reliability is based on scientific validity. (Rehnquist, Ch. J., and Stevens, J., dissented in part from this holding.)

[***LEdHN4]
EVIDENCE §641
expert scientific testimony -- relevance --
Headnote: [4A] [4B]
The "helpfulness" standard of Rule 702 of the Federal Rules of Evidence (FRE), which requires that scientific evidence or testimony assist the trier of fact to understand the evidence or to determine a fact in issue--a condition that goes primarily to relevance--requires a valid scientific connection to the pertinent inquiry

as a precondition to admissibility; for purposes of Rule 702, expert testimony which does not relate to any issue in the case at hand is not relevant and thus is nonhelpful.

[***LEdHN5]
EVIDENCE §641
expert witnesses --
Headnote: [5]
Unlike an ordinary witness, an expert witness is permitted wide latitude, under the Federal Rules of Evidence, to offer opinions, including those that are not based on firsthand knowledge or observation.

[***LEdHN6]
EVIDENCE §641
expert scientific testimony --
Headnote: [6]
Pursuant to Rule 104(a) of the Federal Rules of Evidence, governing preliminary questions concerning the admissibility of evidence, a federal trial judge who is faced with a proffer of expert scientific testimony must determine at the outset whether the expert is proposing to testify to scientific knowledge that will assist the trier of fact to understand or determine a fact in issue; this determination entails a preliminary assessment of (1) whether the reasoning or methodology underlying the testimony is scientifically valid, and (2) whether that reasoning or methodology properly can be applied to the facts in issue. (Rehnquist, Ch. J., and Stevens, J., dissented in part from this holding.)

[***LEdHN7]
EVIDENCE §383
burden of proof --
Headnote: [7A] [7B]
Matters to be determined by a federal trial court pursuant to Rule 104(a) of the Federal Rules of Evidence--that is, preliminary questions concerning the qualification of a person to be a witness, the existence of a privilege, or the admissibility of evidence--are to be established by a preponderance of proof.

[***LEdHN8]
EVIDENCE §67
science -- admissibility -- judicial notice --
Headnote: [8A] [8B]
The requirements of Rule 702 of the Federal Rules of Evidence (FRE) for the admissibility of expert scientific evidence do not apply specially or exclusively to unconventional evidence; however, theories that are so firmly established as to have attained the status of scientific law, such as the laws of thermodynamics, properly are subject to judicial notice under Rule 201 of the FRE.

[***LEdHN9]
EVIDENCE §641

expert scientific testimony -- admissibility --
Headnote: [9]
In determining whether a theory or technique is scientific knowledge that will assist the trier of fact, so as to be the basis of admissible evidence under Rule 702 of the Federal Rules of Evidence, (1) a key question to be answered is, ordinarily, whether the theory or technique can be and has been tested; (2) a pertinent consideration is whether the theory or technique has been subjected to peer review and publication, although the fact of publication, or lack thereof, in a peer-reviewed journal is not a dispositive consideration; (3) the court should ordinarily consider the known or potential rate of error of a particular scientific technique; (4) the assessment of reliability permits, but does not require, explicit identification of a relevant scientific community and an express determination of a particular degree of acceptance of the theory or technique within that community, as (a) widespread acceptance can be an important factor in ruling particular evidence admissible, and (b) a known technique that has been able to attract only minimal support within the scientific community may properly be viewed with skepticism; and (5) the inquiry is a flexible one, and the focus must be solely on principles and methodology, not on the conclusions that such principles and methodology generate. (Rehnquist, Ch. J., and Stevens, J., dissented in part from this holding.)

[***LEdHN10]
TRIAL §15
witnesses -- control --
Headnote: [10]
Since expert evidence can be both powerful and misleading because of the difficulty in evaluating such evidence, a federal trial judge--in weighing, under Rule 403 of the Federal Rules of Evidence, the possible danger of unfair prejudice resulting from such evidence against the evidence's probative force--exercises more control over experts than over lay witnesses.

[***LEdHN11]
EVIDENCE §641
SUMMARY JUDGMENT AND JUDGMENT ON PLEADINGS §1
TRIAL §199
WITNESSES §59
scientific testimony -- attack --
Headnote: [11]
In federal cases, the appropriate means of attacking scientific testimony, where the basis of such testimony meets the admissibility standards of Rule 702 of the Federal Rules of Evidence, are (1) vigorous cross-examination, (2) presentation of contrary evidence, and (3) careful instruction on the burden of proof; additionally, in the event that the trial court concludes that a scintilla of such evidence presented to support a position is insufficient to allow a reasonable juror to conclude that the position more likely than not is true, the court remains free to direct a judgment under Rule 50(a) of the Federal Rules of Civil Procedure (FRCP) and to grant summary judgment under Rule 56 of the FRCP.

[***LEdHN12]
APPEAL §1692.3
remand -- misconception of law --
Headnote: [12]

On certiorari to review a United States Court of Appeals judgment which upheld a United States District Court's ruling that proffered scientific evidence as to the alleged causation of birth defects was inadmissible, the United States Supreme Court will vacate the Court of Appeals' judgment and remand the case for further proceedings, where (1) the inquiries of the District Court and the Court of Appeals as to the admissibility of the evidence focused almost exclusively on whether the principle on which the evidence was based had gained "general acceptance," as gauged by publication and the decisions of other courts; and (2) the Supreme Court holds that general acceptance is not a necessary precondition to the admissibility of scientific evidence under the Federal Rules of Evidence.

SYLLABUS:

Petitioners, two minor children and their parents, alleged in their suit against respondent that the children's serious birth defects had been caused by the mothers' prenatal ingestion of Bendectin, a prescription drug marketed by respondent. The District Court granted respondent summary judgment based on a well-credentialed expert's affidavit concluding, upon reviewing the extensive published scientific literature on the subject, that maternal use of Bendectin has not been shown to be a risk factor for human birth defects. Although petitioners had responded with the testimony of eight other well-credentialed experts, who based their conclusion that Bendectin can cause birth defects on animal studies, chemical structure analyses, and the unpublished "reanalysis" of previously published human statistical studies, the court determined that this evidence did not meet the applicable "general acceptance" standard for the admission of expert testimony. The Court of Appeals agreed and affirmed, citing Frye v. United States, 54 App. D.C. 46, 47, 293 F. 1013, 1014, for the rule that expert opinion based on a scientific technique is inadmissible unless the technique is "generally accepted" as reliable in the relevant scientific community.

Held: The Federal Rules of Evidence, not Frye, provide the standard for admitting expert scientific testimony in a federal trial. Pp. 585-597.

(a) Frye's "general acceptance" test was superseded by the Rules' adoption. The Rules occupy the field, United States v. Abel, 469 U.S. 45, 49, 83 L. Ed. 2d 450, 105 S. Ct. 465, and, although the common law of evidence may serve as an aid to their application, id., at 51-52, respondent's assertion that they some-how assimilated Frye is unconvincing. Nothing in the Rules as a whole or in the text and drafting history of Rule 702, which specifically governs expert testimony, gives any indication that "general acceptance" is a necessary precondition to the admissibility of scientific evidence. Moreover, such a rigid standard would be at odds with the Rules' liberal thrust and their general approach of relaxing the traditional barriers to "opinion" testimony. Pp. 585-589.

(b) The Rules -- especially Rule 702 -- place appropriate limits on the admissibility of purportedly scientific evidence by assigning to the trial judge the task of ensuring that an expert's testimony both rests on a reliable foundation and is relevant to the task at hand. The reliability standard is established by Rule 702's requirement that an expert's testimony pertain to "scientific...knowledge," since the adjective "scientific" implies a grounding in science's methods and procedures, while the word "knowledge" connotes a body of known facts or of ideas inferred from such facts or accepted as true on good grounds. The Rule's requirement that the testimony "assist the trier of fact to understand the evidence or to determine a fact in issue" goes primarily to relevance by demanding a valid scientific connection to the pertinent inquiry as a precondition to admissibility. Pp. 589-592.

(c) Faced with a proffer of expert scientific testimony under Rule 702, the trial judge, pursuant to Rule 104(a), must make a preliminary assessment of whether the testimony's underlying reasoning or methodology is scientifically valid and properly can be applied to the facts at issue. Many considerations will bear on the inquiry, including whether the theory or technique in question can be (and has been) tested, whether it has been subjected to peer review and publication, its known or potential error rate and the existence and maintenance of standards controlling its operation, and whether it has attracted widespread acceptance within a relevant scientific community. The inquiry is a flexible one, and its focus must be solely on principles and methodology, not on the conclusions that they generate. Throughout, the judge should also be mindful of other applicable Rules. Pp. 592-595.

(d) Cross-examination, presentation of contrary evidence, and careful instruction on the burden of proof, rather than wholesale exclusion under an uncompromising "general acceptance" standard, is the appropriate means by which evidence based on valid principles may be challenged. That even limited screening by the trial judge, on occasion, will prevent the jury from hearing of authentic scientific breakthroughs is simply a consequence of the fact that the Rules are not designed to seek cosmic understanding but, rather, to resolve legal disputes. Pp. 595-597.

COUNSEL: Michael H. Gottesman argued the cause for petitioners. With him on the briefs were Kenneth J. Chesebro, Barry J. Nace, David L. Shapiro, and Mary G. Gillick.

Charles Fried argued the cause for respondent. With him on the brief were Charles R. Nesson, Joel I. Klein, Richard G. Taranto, Hall R. Marston, George E. Berry, Edward H. Stratemeier, and W. Glenn Forrester. *

* Briefs of amici curiae urging reversal were filed for the State of Texas et al. by Dan Morales, Attorney General of Texas, Mark Barnett, Attorney General of South Dakota, Marc Racicot, Attorney General of Montana, Larry EchoHawk, Attorney General of Idaho, and Brian Stuart Koukoutchos; for the American Society of Law, Medicine and Ethics et al. by Joan E. Bertin, Marsha S. Berzon, and Albert H. Meyerhoff; for the Association of Trial Lawyers of America by Jeffrey Robert White and Roxanne Barton Conlin; for Ronald Bayer et al. by Brian Stuart Koukoutchos, Priscilla Budeiri, Arthur Bryant, and George W. Conk; and for Daryl E. Chubin et al. by Ron Simon and Nicole Schultheis.

Briefs of amici curiae urging affirmance were filed for the United States by Acting Solicitor General Wallace, Assistant Attorney General Gerson, Miguel A.

Estrada, Michael Jay Singer, and John P. Schnitker; for the American Insurance Association by William J. Kilberg, Paul Blankenstein, Bradford R. Clark, and Craig A. Berrington; for the American Medical Association et al. by Carter G. Phillips, Mark D. Hopson, and Jack R. Bierig; for the American Tort Reform Association by John G. Kester and John W. Vardaman, Jr.; for the Chamber of Commerce of the United States by Timothy B. Dyk, Stephen A. Bokat, and Robin S. Conrad; for the Pharmaceutical Manufacturers Association by Louis R. Cohen and Daniel Marcus; for the Product Liability Advisory Council, Inc., et al. by Victor E. Schwartz, Robert P. Charrow, and Paul F. Rothstein; for the Washington Legal Foundation by Scott G. Campbell, Daniel J. Popeo, and Richard A. Samp; and for Nicolaas Bloembergen et al. by Martin S. Kaufman.

Briefs of amici curiae were filed for the American Association for the Advancement of Science et al. by Richard A. Meserve and Bert Black; for the American College of Legal Medicine by Miles J. Zaremski; for the Carnegie Commission on Science, Technology, and Government by Steven G. Gallagher, Elizabeth H. Esty, and Margaret A. Berger; for the Defense Research Institute, Inc., by Joseph A. Sherman, E. Wayne Taff, and Harvey L. Kaplan; for the New England Journal of Medicine et al. by Michael Malina and Jeffrey I. D. Lewis; for A Group of American Law Professors by Donald N. Bersoff; for Alvan R. Feinstein by Don M. Kennedy, Loretta M. Smith, and Richard A. Oetheimer; and for Kenneth Rothman et al. by Neil B. Cohen.

JUDGES: BLACKMUN, J., delivered the opinion for a unanimous Court with respect to Parts I and II-A, and the opinion of the Court with respect to Parts II-B, II-C, III, and IV, in which WHITE, O'CONNOR, SCALIA, KENNEDY, SOUTER, and THOMAS, JJ., joined. REHNQUIST, C. J., filed an opinion concurring in part and dissenting in part, in which STEVENS, J., joined, post, p. 598.

OPINIONBY: BLACKMUN

OPINION:

[*582] [***476] [**2791] JUSTICE BLACKMUN delivered the opinion of the Court.

[***LEdHR1A] [1A]In this case we are called upon to determine the standard for admitting expert scientific testimony in a federal trial.

I

Petitioners Jason Daubert and Eric Schuller are minor children born with serious birth defects. They and their parents sued respondent in California state court, alleging that the birth defects had been caused by the mothers' ingestion of Bendectin, a prescription antinausea drug marketed by respondent. Respondent removed the suits to federal court on diversity grounds.

After extensive discovery, respondent moved for summary judgment, contending that Bendectin does not cause birth defects in humans and that petitioners would be unable to come forward with any admissible evidence that it does. In support of its motion, respondent submitted an affidavit of Steven H. Lamm, physician and epidemiologist, who is a well-credentialed expert on the risks from exposure to various chemical substances. n1 Doctor Lamm stated that he had reviewed all the literature on Bendectin and human birth defects -- more than 30 published studies involving over 130,000 patients. No study had found Bendectin to be a human teratogen (i.e., a substance capable of causing malformations in fetuses). On the basis of this review, Doctor Lamm concluded that maternal use of Bendectin during the first trimester of pregnancy has not been shown to be a risk factor for human birth defects.

- - - - - - - - - - - - - - Footnotes - - - - - - - - - - - - - - -

n1 Doctor Lamm received his master's and doctor of medicine degrees from the University of Southern California. He has served as a consultant in birth-defect epidemiology for the National Center for Health Statistics and has published numerous articles on the magnitude of risk from exposure to various chemical and biological substances. App. 34-44.

- - - - - - - - - - - - End Footnotes- - - - - - - - - - - - - -

[*583] Petitioners did not (and do not) contest this characterization of the published record regarding Bendectin. Instead, they responded to respondent's motion with the testimony of eight experts of their own, each of whom also possessed impressive credentials. n2 These experts had concluded that Bendectin can cause birth defects. Their conclusions were based upon "in vitro" (test tube) and "in vivo" (live) animal studies that [***477] found a link between Bendectin and malformations; pharmacological studies of the chemical structure of Bendectin that purported to show similarities between the structure of the drug and that of other substances known to cause birth defects; and the "reanalysis" of previously [**2792] published epidemiological (human statistical) studies.

- - - - - - - - - - - - - - Footnotes - - - - - - - - - - - - - - -

n2 For example, Shanna Helen Swan, who received a master's degree in biostatistics from Columbia University and a doctorate in statistics from the University of California at Berkeley, is chief of the section of the California Department of Health and Services that determines causes of birth defects and has served as a consultant to the World Health Organization, the Food and Drug Administration, and the National Institutes of Health. Id., at 113-114, 131-132. Stuart A. Newman, who received his bachelor's degree in chemistry from Columbia University and his master's and doctorate in chemistry from the University of Chicago, is a professor at New York Medical College and has spent over a decade studying the effect of chemicals on limb development. Id., at 54-56. The credentials of the others are similarly impressive. See id., at 61-66, 73-80, 148-153, 187-192, and Attachments 12, 20, 21, 26, 31, and 32 to Petitioners' Opposition to Summary Judgment in No. 84-2013-G(I) (SD Cal.).

- - - - - - - - - - - - End Footnotes- - - - - - - - - - - - - -

The District Court granted respondent's motion for summary judgment. The court stated that scientific evidence is admissible only if the principle upon which it is based is "'sufficiently established to have general acceptance in the field to which it belongs.'" 727 F. Supp. 570, 572 (SD Cal. 1989), quoting United States v. Kilgus, 571 F.2d 508, 510 (CA9 1978). The court concluded that petitioners' evidence did not meet this standard. Given the vast body of epidemiological data concerning Bendectin, the court held, expert opinion which is not based on epidemiological evidence [*584] is not admissible to establish causation. 727 F. Supp. at 575. Thus, the animal-cell studies, live-animal studies, and chemical-structure analyses on which petitioners had relied could not raise by themselves a reasonably disputable jury issue regarding causation. Ibid. Petitioners' epidemiological analyses, based as they were on recalculations of data in previously published studies that had found no causal link between the drug and birth defects, were ruled to be inadmissible because they had not been published or subjected to peer review. Ibid.

The United States Court of Appeals for the Ninth Circuit affirmed. 951 F.2d 1128 (1991). Citing Frye v. United States, 54 App. D.C. 46, 47, 293 F. 1013, 1014 (1923), the court stated that expert opinion based on a scientific technique is inadmissible unless the technique is "generally accepted" as reliable in the relevant scientific community. 951 F.2d at 1129-1130. The court declared that expert opinion based on a methodology that diverges "significantly from the procedures accepted by recognized authorities in the field . . . cannot be shown to be 'generally accepted as a reliable technique.'" Id., at 1130, quoting United States v. Solomon, 753 F.2d 1522, 1526 (CA9 1985).

The court emphasized that other Courts of Appeals considering the risks of Bendectin had refused to admit reanalyses of epidemiological studies that had been neither published nor subjected to peer review. 951 F.2d at 1130-1131. Those courts had found unpublished reanalyses "particularly problematic in light of the massive weight of the original published studies supporting [respondent's] position, all of which had undergone full scrutiny from the scientific community". Id., at 1130. Contending that reanalysis is generally accepted by the scientific community only when it is subjected to verification and scrutiny by others in the field, the Court of Appeals rejected petitioners' reanalyses as "unpublished, not subjected to the normal peer review process and generated solely for use in litigation". Id., at 1131. The [*585] court concluded that petitioners' evidence provided an insufficient foundation to allow admission of expert testimony that Bendectin caused their injuries and, accordingly, that petitioners could not satisfy their burden of proving causation at trial.

We granted certiorari, 506 U.S. 914 [***478] (1992), in light of sharp divisions among the courts regarding the proper standard for the admission of expert testimony. Compare, e.g., United States v. Shorter, 257 U.S. App. D.C. 358, 363-364, 809 F.2d 54, 59-60 (applying the "general acceptance" standard), cert. denied, 484 U.S. 817, 98 L. Ed. 2d 35, 108 S. Ct. 71 (1987), with DeLuca v. Merrell Dow Pharmaceuticals, Inc., 911 F.2d 941, 955 (CA3 1990) (rejecting the "general acceptance" standard).

II

A

In the 70 years since its formulation in the Frye case, the "general acceptance" test has been the dominant standard for determining the admissibility of novel scientific evidence at trial. See E. Green & C. Nesson, Problems, Cases, and Materials on Evidence 649 (1983). Although under increasing attack of late, the rule continues to be followed by a [**2793] majority of courts, including the Ninth Circuit. n3

- - - - - - - - - - - - - - Footnotes - - - - - - - - - - - - - - -

n3 For a catalog of the many cases on either side of this controversy, see P. Giannelli & E. Imwinkelried, Scientific Evidence § 1-5, pp. 10-14 (1986 and Supp. 1991).

- - - - - - - - - - - - End Footnotes- - - - - - - - - - - - - -

The Frye test has its origin in a short and citation-free 1923 decision concerning the admissibility of evidence derived from a systolic blood pressure deception test, a crude precursor to the polygraph machine. In what has become a famous (perhaps infamous) passage, the then Court of Appeals for the District of Columbia described the device and its operation and declared:

"Just when a scientific principle or discovery crosses the line between the experimental and demonstrable stages [*586] is difficult to define. Somewhere in this twilight zone the evidential force of the principle must be recognized, and while courts will go a long way in admitting expert testimony deduced from a well-recognized scientific principle or discovery, the thing from which the deduction is made must be sufficiently established to have gained general acceptance in the particular field in which it belongs". 54 App. D.C. at 47, 293 F. at 1014 (emphasis added).

Because the deception test had "not yet gained such standing and scientific recognition among physiological and psychological authorities as would justify the courts in admitting expert testimony deduced from the discovery, development, and experiments thus far made," evidence of its results was ruled inadmissible. Ibid.

[***LEdHR1B] [1B]The merits of the Frye test have been much debated, and scholarship on its proper scope and application is legion. n4 [*587] Petitioners' primary [***479] attack, however, is not on the content but on the continuing authority of the rule. They contend that the Frye test was superseded by the adoption of the Federal Rules of Evidence. n5 We agree.

- - - - - - - - - - - - - - Footnotes - - - - - - - - - - - - - - -

n4 See, e.g., Green, Expert Witnesses and Sufficiency of Evidence in Toxic Substances Litigation: The Legacy of Agent Orange and Bendectin Litigation, 86 Nw. U. L. Rev. 643 (1992) (hereinafter Green); Becker & Orenstein, The Federal Rules of Evidence After Sixteen Years -- The Effect of "Plain Meaning" Jurisprudence, the Need for an Advisory Committee on the Rules of Evidence, and Suggestions for Selective Revision of the Rules, 60 Geo. Wash. L. Rev. 857, 876-885 (1992); Hanson, James Alphonzo Frye is Sixty-Five Years Old; Should He Retire?, 16 West. St. U. L. Rev. 357 (1989); Black, A Unified Theory of Scientific Evidence, 56 Ford. L. Rev. 595 (1988); Imwinkelried, The "Bases" of Expert

Testimony: The Syllogistic Structure of Scientific Testimony, 67 N. C. L. Rev. 1 (1988); Proposals for a Model Rule on the Admissibility of Scientific Evidence, 26 Jurimetrics J. 235 (1986); Giannelli, The Admissibility of Novel Scientific Evidence: Frye v. United States, a Half-Century Later, 80 Colum. L. Rev. 1197 (1980); The Supreme Court, 1986 Term, 101 Harv. L. Rev. 7, 119, 125-127 (1987).

Indeed, the debates over Frye are such a well-established part of the academic landscape that a distinct term -- "Frye-ologist" -- has been advanced to describe those who take part. See Behringer, Introduction, Proposals for a Model Rule on the Admissibility of Scientific Evidence, 26 Jurimetrics J. 237, 239 (1986), quoting Lacey, Scientific Evidence, 24 Jurimetrics J. 254, 264 (1984).

n5 Like the question of Frye's merit, the dispute over its survival has divided courts and commentators. Compare, e.g., United States v. Williams, 583 F.2d 1194 (CA2 1978) (Frye is superseded by the Rules of Evidence), cert. denied, 439 U.S. 1117, 59 L. Ed. 2d 77, 99 S. Ct. 1025 (1979), with Christophersen v. Allied-Signal Corp., 939 F.2d 1106, 1111, 1115-1116 (CA5 1991) (en banc) (Frye and the Rules coexist), cert. denied, 503 U.S. 912, 117 L. Ed. 2d 506, 112 S. Ct. 1280 (1992), 3 J. Weinstein & M. Berger, Weinstein's Evidence P702[03], pp. 702-36 to 702-37 (1988) (hereinafter Weinstein & Berger) (Frye is dead), and M. Graham, Handbook of Federal Evidence § 703.2 (3d ed. 1991) (Frye lives). See generally P. Giannelli & E. Imwinkelried, Scientific Evidence § 1-5, at 28-29 (citing authorities).

------------ End Footnotes- - - - - - - - - - - - -

[***LEdHR2] [2]We interpret the legislatively enacted Federal Rules of Evidence as we would any statute. Beech Aircraft Corp. v. Rainey, 488 U.S. 153, 163, 102 L. Ed. 2d 445, 109 S. Ct. 439 (1988). Rule 402 provides the baseline:

"All relevant evidence is admissible, except as otherwise provided by the Constitution of the United States, by Act of Congress, [**2794] by these rules, or by other rules prescribed by the Supreme Court pursuant to statutory authority. Evidence which is not relevant is not admissible".

"Relevant evidence" is defined as that which has "any tendency to make the existence of any fact that is of consequence to the determination of the action more probable or less probable than it would be without the evidence". Rule 401. The Rules' basic standard of relevance thus is a liberal one.

Frye, of course, predated the Rules by half a century. In United States v. Abel, 469 U.S. 45, 83 L. Ed. 2d 450, 105 S. Ct. 465 (1984), we considered the pertinence of background common law in interpreting the Rules of Evidence. We noted that the Rules occupy the field, id., at 49, but, quoting Professor Cleary, the Reporter, [*588] explained that the common law nevertheless could serve as an aid to their application:

"'In principle, under the Federal Rules no common law of evidence remains. "All relevant evidence is admissible, except as otherwise provided . . .". In reality, of course, the body of common law knowledge continues to exist, though in the somewhat altered form of a source of guidance in the exercise of delegated powers.'" Id., at 51-52.

We found the common-law precept at issue in the Abel case entirely consistent with Rule 402's general requirement of admissibility, and considered it unlikely that the drafters had intended to change the rule. Id., at 50-51. In Bourjaily v. United States, 483 U.S. 171, 97 L. Ed. 2d 144, 107 S. [***480] Ct. 2775 (1987), on the other hand, the Court was unable to find a particular common-law doctrine in the Rules, and so held it superseded.

[***LEdHR1C] [1C] Here there is a specific Rule that speaks to the contested issue. Rule 702, governing expert testimony, provides:

"If scientific, technical, or other specialized knowledge will assist the trier of fact to understand the evidence or to determine a fact in issue, a witness qualified as an expert by knowledge, skill, experience, training, or education, may testify thereto in the form of an opinion or otherwise".

Nothing in the text of this Rule establishes "general acceptance" as an absolute prerequisite to admissibility. Nor does respondent present any clear indication that Rule 702 or the Rules as a whole were intended to incorporate a "general acceptance" standard. The drafting history makes no mention of Frye, and a rigid "general acceptance" requirement would be at odds with the "liberal thrust" of the Federal Rules and their "general approach of relaxing the traditional barriers to 'opinion' testimony". Beech Aircraft Corp. v. Rainey, 488 U.S. at 169 (citing Rules 701 to 705). See also Weinstein, Rule 702 of the Federal Rules of Evidence is [*589] Sound; It Should Not Be Amended, 138 F.R.D. 631 (1991) ("The Rules were designed to depend primarily upon lawyer-adversaries and sensible triers of fact to evaluate conflicts"). Given the Rules' permissive backdrop and their inclusion of a specific rule on expert testimony that does not mention "general acceptance," the assertion that the Rules somehow assimilated Frye is unconvincing. Frye made "general acceptance" the exclusive test for admitting expert scientific testimony. That austere standard, absent from, and incompatible with, the Federal Rules of Evidence, should not be applied in federal trials. n6

[***LEdHR1D] [1D]

- - - - - - - - - - - - - - Footnotes - - - - - - - - - - - - - -

n6 Because we hold that Frye has been superseded and base the discussion that follows on the content of the congressionally enacted Federal Rules of Evidence, we do not address petitioners' argument that application of the Frye rule in this diversity case, as the application of a judgemade rule affecting substantive rights, would violate the doctrine of Erie R. Co. v. Tompkins, 304 U.S. 64, 82 L. Ed. 1188, 58 S. Ct. 817 (1938).

- - - - - - - - - - - - End Footnotes- - - - - - - - - - - - -

B

[***LEdHR3A] [3A]That the Frye test was displaced by the Rules of Evidence does not mean, [**2795] however, that the Rules themselves place no limits on the admissibility of purportedly scientific evidence. n7 Nor is the trial judge disabled from screening such evidence. To the contrary, under the Rules the trial judge must ensure that any and all scientific testimony or evidence admitted is not only relevant, but reliable.

- - - - - - - - - - - - - - Footnotes - - - - - - - - - - - - - -

n7 THE CHIEF JUSTICE "do[es] not doubt that Rule 702 confides to the judge some gatekeeping responsibility," post, at 600, but would neither say how it does so nor explain what that role entails. We believe the better course is to note the nature and source of the duty.

- - - - - - - - - - - - End Footnotes- - - - - - - - - - - - -

The primary locus of this obligation is Rule 702, which clearly contemplates some degree of regulation of the subjects and theories about which an expert may testify. "If scientific, technical, or other specialized knowledge will assist the trier of fact to understand the evidence or to determine a fact in issue" an expert "may testify thereto". (Emphasis added.) The subject of an expert's testimony must [*590] be "scientific . . . [***481] knowledge". n8 The adjective "scientific" implies a grounding in the methods and procedures of science. Similarly, the word "knowledge" connotes more than subjective belief or unsupported speculation. The term "applies to any body of known facts or to any body of ideas inferred from such facts or accepted as truths on good grounds". Webster's Third New International Dictionary 1252 (1986). Of course, it would be unreasonable to conclude that the subject of scientific testimony must be "known" to a certainty; arguably, there are no certainties in science. See, e.g., Brief for Nicolaas Bloembergen et al. as Amici Curiae 9 ("Indeed, scientists do not assert that they know what is immutably 'true' -- they are committed to searching for new, temporary, theories to explain, as best they can, phenomena"); Brief for American Association for the Advancement of Science et al. as Amici Curiae 7-8 ("Science is not an encyclopedic body of knowledge about the universe. Instead, it represents a process for proposing and refining theoretical explanations about the world that are subject to further testing and refinement" (emphasis in original)). But, in order to qualify as "scientific knowledge," an inference or assertion must be derived by the scientific method. Proposed testimony must be supported by appropriate validation -- i.e., "good grounds," based on what is known. In short, the requirement that an expert's testimony pertain to "scientific knowledge" establishes a standard of evidentiary reliability. n9

- - - - - - - - - - - - - - Footnotes - - - - - - - - - - - - - -

n8 Rule 702 also applies to "technical, or other specialized knowledge". Our discussion is limited to the scientific context because that is the nature of the expertise offered here.

[***LEdHR3B] [3B]

n9 We note that scientists typically distinguish between "validity" (does the principle support what it purports to show?) and "reliability" (does application of the principle produce consistent results?). See Black, 56 Ford. L. Rev., at 599. Although "the difference between accuracy, validity, and reliability may be such that each is distinct from the other by no more than a hen's kick," Starrs, Frye v. United States Restructured and Revitalized: A Proposal to Amend Federal Evidence Rule 702, 26 Jurimetrics J. 249, 256 (1986), our reference here is to evidentiary reliability -- that is, trustworthiness. Cf., e.g., Advisory Committee's Notes on Fed. Rule Evid. 602, 28 U.S.C. App., p. 755 ("'The rule requiring that a witness who testifies to a fact which can be perceived by the senses must have had an opportunity to observe, and must have actually observed the fact' is a 'most pervasive manifestation' of the common law insistence upon 'the most reliable sources of information'" (citation omitted)); Advisory Committee's Notes on Art. VIII of Rules of Evidence, 28 U.S.C. App., p. 770 (hearsay exceptions will be recognized only "under circumstances supposed to furnish guarantees of trustworthiness"). In a case involving scientific evidence, evidentiary reliability will be based upon scientific validity.

- - - - - - - - - - - - End Footnotes- - - - - - - - - - - - - -

[*591]

[***LEdHR4A] [4A]Rule 702 further requires that the evidence or testimony "assist the trier of fact to understand the evidence or to determine a fact in issue". This condition goes primarily to relevance. "Expert testimony which does not relate to any issue in the case is not relevant and, ergo, non-helpful". 3 Weinstein & Berger P702[02], p. 702-18. See also United States v. Downing, 753 F.2d 1224, 1242 (CA3 1985) ("An additional consideration [**2796] under Rule 702 -- and another aspect of relevancy -- is whether expert testimony proffered in the case is sufficiently tied to the facts of the case that it will aid the jury in resolving a factual dispute"). The consideration has been aptly described by Judge Becker as one of "fit". Ibid. "Fit" is not always obvious, [***482] and scientific validity for one purpose is not necessarily scientific validity for other, unrelated purposes. See Starrs, Frye v. United States Restructured and Revitalized: A Proposal to Amend Federal Evidence Rule 702, 26 Jurimetrics J. 249, 258 (1986). The study of the phases of the moon, for example, may provide valid scientific "knowledge" about whether a certain night was dark, and if darkness is a fact in issue, the knowledge will assist the trier of fact. However (absent creditable grounds supporting such a link), evidence that the moon was full on a certain night will not assist the trier of fact in determining whether an individual was unusually likely to have behaved irrationally on that night. Rule 702's "helpfulness" [*592] standard requires a valid scientific connection to the pertinent inquiry as a precondition to admissibility.

[***LEdHR5] [5]That these requirements are embodied in Rule 702 is not surprising. Unlike an ordinary witness, see Rule 701, an expert is permitted wide latitude to offer opinions, including those that are not based on firsthand knowledge or observation. See Rules 702 and 703. Presumably, this relaxation of the usual requirement of firsthand knowledge -- a rule which represents "a 'most pervasive manifestation' of the common law insistence upon 'the most reliable sources of information,'" Advisory Committee's Notes on Fed. Rule Evid. 602, 28 U.S.C. App.,

p. 755 (citation omitted) -- is premised on an assumption that the expert's opinion will have a reliable basis in the knowledge and experience of his discipline.

C

[***LEdHR6] [6] [***LEdHR7A] [7A] [***LEdHR8A] [8A]Faced with a proffer of expert scientific testimony, then, the trial judge must determine at the outset, pursuant to Rule 104(a), n10 whether the expert is proposing to testify to (1) scientific knowledge that (2) will assist the trier of fact to understand or determine a fact in issue. n11 This entails a preliminary assessment of whether the reasoning or methodology [*593] underlying the testimony is scientifically valid and of whether that reasoning or methodology properly can be applied to the facts in issue. We are confident that federal judges possess the capacity to undertake this review. Many factors will bear on the inquiry, and we do not presume to set out a definitive checklist or test. But some general observations are appropriate.

[***LEdHR7B] [7B]

- - - - - - - - - - - - - - Footnotes - - - - - - - - - - - - - - -

n10 Rule 104(a) provides:

"Preliminary questions concerning the qualification of a person to be a witness, the existence of a privilege, or the admissibility of evidence shall be determined by the court, subject to the provisions of subdivision (b) [pertaining to conditional admissions]. In making its determination it is not bound by the rules of evidence except those with respect to privileges". These matters should be established by a preponderance of proof. See Bourjaily v. United States, 483 U.S. 171, 175-176, 97 L. Ed. 2d 144, 107 S. Ct. 2775 (1987).

[***LEdHR8B] [8B]

n11 Although the Frye decision itself focused exclusively on "novel" scientific techniques, we do not read the requirements of Rule 702 to apply specially or exclusively to unconventional evidence. Of course, well-established propositions are less likely to be challenged than those that are novel, and they are more handily defended. Indeed, theories that are so firmly established as to have attained the status of scientific law, such as the laws of thermodynamics, properly are subject to judicial notice under Federal Rule of Evidence 201.

- - - - - - - - - - - - End Footnotes- - - - - - - - - - - - - -

[***LEdHR9] [9]Ordinarily, a key question to be answered in determining whether a theory or technique is scientific knowledge that will assist the trier of fact will be whether it can be (and has been) tested. "Scientific methodology today is based on generating hypotheses and testing them to see if they can be falsified; indeed, this methodology is what distinguishes science from other fields of human inquiry". Green 645. See also C. Hempel, Philosophy of Natural Science 49 (1966) ("The statements constituting a scientific explanation must be capable of empirical test"); K. Popper, Conjectures and Refutations: The Growth of Scientific Knowledge 37 (5th ed. 1989) ("The criterion of the scientific status of a theory is its falsifiability, or refutability, or testability") (emphasis deleted).

Another pertinent consideration is whether the theory or technique has been subjected to peer review and publication. Publication (which is but one element of peer review) is not a sine qua non of admissibility; it does not necessarily correlate with reliability, see S. Jasanoff, The Fifth Branch: Science Advisors as Policymakers 61-76 (1990), and in some instances well-grounded but innovative theories will not have been published, see Horrobin, The Philosophical Basis of Peer Review and the Suppression of Innovation, 263 JAMA 1438 (1990). Some propositions, moreover, are too particular, too new, or of too limited interest to be published. But submission to the scrutiny of the scientific community is a component of "good science," in part because it increases the likelihood that substantive flaws in methodology will be detected. See J. Ziman, Reliable Knowledge: An Exploration of the Grounds for Belief in Science 130-133 (1978); Relman & Angell, How Good Is Peer Review?, 321 New Eng. J. Med. 827 (1989). The fact of publication (or lack thereof) in a peer reviewed journal thus will be a relevant, though not dispositive, consideration in assessing the scientific validity of a particular technique or methodology on which an opinion is premised.

Additionally, in the case of a particular scientific technique, the court ordinarily should consider the known or potential rate of error, see, e.g., United States v. Smith, 869 F.2d 348, 353-354 (CA7 1989) (surveying studies of the error rate of spectrographic voice identification technique), and the existence and maintenance of standards controlling the technique's operation, see United States v. Williams, 583 F.2d 1194, 1198 (CA2 1978) (noting professional organization's standard governing spectrographic analysis), cert. denied, 439 U.S. 1117, 59 L. Ed. 2d 77, 99 S. Ct. 1025 (1979).

Finally, "general acceptance" can yet have a bearing on the inquiry. A "reliability assessment does not require, although it does permit, explicit identification of a relevant scientific community and an express determination of a particular degree of acceptance within that community". United States v. Downing, 753 F.2d at 1238. See also 3 Weinstein & Berger P702[03], pp. 702-41 to 702-42. Widespread acceptance can be an important factor in ruling particular evidence admissible, and "a known technique which has been able to attract only minimal support within the community," Downing, 753 F.2d at 1238, may properly be viewed with skepticism.

The inquiry envisioned by Rule 702 is, we emphasize, a flexible one. n12 Its overarching subject is the scientific validity [*595] -- and thus the evidentiary relevance and reliability -- of the principles that underlie a proposed submission. The focus, of course, must be solely on principles and methodology, not on the conclusions that they generate.

-------------- Footnotes ---------------

n12 A number of authorities have presented variations on the reliability approach, each with its own slightly different set of factors. See, e.g., Downing, 753 F.2d at 1238-1239 (on which our discussion draws in part); 3 Weinstein & Berger P702[03], pp. 702-41 to 702-42 (on which the Downing court in turn partially relied); McCormick, Scientific Evidence: Defining a New Approach to Admissibility, 67 Iowa L. Rev. 879, 911-912 (1982); and Symposium on Science and the Rules of Evidence, 99 F.R.D. 187, 231 (1983) (statement by Margaret Berger). To the extent that they focus on the reliability of evidence as ensured by the scientific validity of its underlying principles, all these versions may well have merit, although we express no opinion regarding any of their particular details.

------------ End Footnotes- - - - - - - - - - - - -

[***LEdHR10] [10]Throughout, a judge assessing a proffer of expert scientific testimony under Rule 702 should also be mindful of other applicable rules. Rule 703 provides that expert opinions based on otherwise inadmissible hearsay are to be admitted only if the facts or data are "of a type reasonably relied upon by experts in the particular field in forming opinions or inferences upon the subject". Rule 706 allows the court at its discretion to procure the assistance of an expert of its own choosing. Finally, Rule 403 permits the exclusion of relevant evidence "if its probative value is substantially outweighed by the danger of unfair prejudice, confusion of the issues, or misleading the jury . . .". Judge Weinstein has explained: "Expert evidence can be both powerful and quite misleading because of the difficulty in evaluating it. Because of this risk, the judge in weighing possible prejudice against probative force under Rule 403 of the present rules exercises more control over experts than over lay witnesses". Weinstein, 138 F.R.D. at 632.

III

[***LEdHR11] [11] We conclude by briefly addressing what appear to be two underlying concerns of the parties and amici in this case. Respondent expresses apprehension that abandonment of "general acceptance" as the exclusive requirement for admission will result in a "free-for-all" in which befuddled juries are confounded by absurd and irrational pseudoscientific assertions.

[*596] In this regard respondent seems to us to be overly pessimistic about the capabilities of the jury and of the adversary system generally. Vigorous cross-examination, presentation of contrary evidence, and careful instruction on the burden of proof are the traditional and appropriate means of attacking shaky but admissible evidence. See Rock v. Arkansas, 483 U.S. 44, 61, 97 L. Ed. 2d 37, 107 S. Ct. 2704 (1987). Additionally, in the event the trial court concludes that the scintilla of evidence presented supporting a position is insufficient to allow a reasonable juror to conclude that the position more likely than not is true, the court remains free to direct a judgment, Fed. Rule Civ. Proc. 50(a), and likewise to grant summary judgment, Fed. Rule Civ. Proc. 56. Cf., e.g., Turpin v. Merrell Dow Pharmaceuticals, Inc., 959 F.2d 1349 (CA6) (holding that scientific evidence that provided foundation for expert testimony, viewed in the light most favorable to plaintiffs, was not sufficient to allow a jury to find it more probable than not that defendant [***485] caused plaintiff's injury), cert. denied, 506 U.S. 826, 121 L. Ed. 2d 47, 113 S. Ct. 84 (1992); Brock v. Merrell Dow Pharmaceuticals, Inc., 874 F.2d 307 (CA5 1989) (reversing judgment entered on jury verdict for plaintiffs because evidence regarding causation was insufficient), modified, 884 F.2d 166 (CA5 1989),

cert. denied, 494 U.S. 1046 (1990); Green 680-681. These conventional devices, rather than wholesale exclusion under an uncompromising "general acceptance" test, are the appropriate safeguards where the basis of scientific testimony meets the standards of Rule 702.

Petitioners and, to a greater extent, their amici exhibit a different concern. They suggest that recognition of a screening role for the judge that allows for the exclusion of "invalid" evidence will sanction a stifling and repressive scientific orthodoxy and will be inimical to the search for truth. See, e.g., Brief for Ronald Bayer et al. as Amici Curiae. It is true that open debate is an essential part of both legal and scientific analyses. Yet there are important differences between the quest for truth in the courtroom and the quest [*597] for truth in the laboratory. Scientific conclusions are subject to perpetual revision. Law, on the other hand, must resolve disputes finally and quickly. The scientific project is advanced by broad and wide-ranging consideration of a multitude of hypotheses, for those that are incorrect will eventually be shown to be so, and that in itself is an advance. Conjectures that are probably wrong are of little use, however, in the project of reaching a quick, final, and binding legal judgment -- often of great consequence – about a particular set of events in the past. We recognize that, in practice, a gatekeeping role for the judge, no matter how flexible, inevitably on occasion will prevent the jury from learning of authentic [**2799] insights and innovations. That, nevertheless, is the balance that is struck by Rules of Evidence designed not for the exhaustive search for cosmic understanding but for the particularized resolution of legal disputes. n13

- - - - - - - - - - - - - - Footnotes - - - - - - - - - - - - - -

n13 This is not to say that judicial interpretation, as opposed to adjudicative factfinding, does not share basic characteristics of the scientific endeavor: "The work of a judge is in one sense enduring and in another ephemeral. . . . In the endless process of testing and retesting, there is a constant rejection of the dross and a constant retention of whatever is pure and sound and fine". B. Cardozo, The Nature of the Judicial Process 178-179 (1921).

- - - - - - - - - - - - End Footnotes- - - - - - - - - - - - -

IV

[***LEdHR1E] [1E] [***LEdHR3C] [3C] [***LEdHR4B] [4B] To summarize: "General acceptance" is not a necessary precondition to the admissibility of scientific evidence under the Federal Rules of Evidence, but the Rules of Evidence -- especially Rule 702 -- do assign to the trial judge the task of ensuring that an expert's testimony both rests on a reliable foundation and is relevant to the task at hand. Pertinent evidence based on scientifically valid principles will satisfy those demands.

[***LEdHR12] [12]The inquiries of the District Court and the Court of Appeals focused almost exclusively on "general acceptance," as gauged by publication and the decisions of other courts. Accordingly, [*598] the judgment of the Court of Appeals is vacated, and the case is remanded for further proceedings consistent with this opinion.

It is so ordered.

CONCURBY: REHNQUIST (In Part)
DISSENTBY: REHNQUIST (In Part)
DISSENT:

[***486] CHIEF JUSTICE REHNQUIST, with whom JUSTICE STEVENS joins, concurring in part and dissenting in part.

The petition for certiorari in this case presents two questions: first, whether the rule of Frye v. United States, 54 App. D.C. 46, 293 F. 1013 (1923), remains good law after the enactment of the Federal Rules of Evidence; and second, if Frye remains valid, whether it requires expert scientific testimony to have been subjected to a peer review process in order to be admissible. The Court concludes, correctly in my view, that the Frye rule did not survive the enactment of the Federal Rules of Evidence, and I therefore join Parts I and II-A of its opinion. The second question presented in the petition for certiorari necessarily is mooted by this holding, but the Court nonetheless proceeds to construe Rules 702 and 703 very much in the abstract, and then offers some "general observations". Ante, at 593.

"General observations" by this Court customarily carry great weight with lower federal courts, but the ones offered here suffer from the flaw common to most such observations -- they are not applied to deciding whether particular testimony was or was not admissible, and therefore they tend to be not only general, but vague and abstract. This is particularly unfortunate in a case such as this, where the ultimate legal question depends on an appreciation of one or more bodies of knowledge not judicially noticeable, and subject to different interpretations in the briefs of the parties and their amici. Twenty-two amicus briefs have been filed in the case, and indeed the Court's opinion contains no fewer than 37 citations to amicus briefs and other secondary sources.

[*599] The various briefs filed in this case are markedly different from typical briefs, in that large parts of them do not deal with decided cases or statutory language -- the sort of material we customarily interpret. Instead, they deal with definitions of scientific knowledge, scientific method, scientific validity, and peer review -- in short, matters far afield from the expertise of judges. This is not to say that such materials are not useful or even necessary in deciding how Rule 702 should be applied; but it is to say that the unusual subject matter should cause us to proceed with great caution in deciding more than we have to, because our reach can so easily exceed our grasp.

But even if it were desirable to make "general observations" not necessary to decide [**2800] the questions presented, I cannot subscribe to some of the observations made by the Court. In Part II-B, the Court concludes that reliability and relevancy are the touchstones of the admissibility of expert testimony. Ante, at 590-592. Federal Rule of Evidence 402 provides, as the Court points out, that "evidence which is not relevant is not admissible". But there is no similar reference in the Rule to "reliability". The Court constructs its argument by parsing the language "if scientific, technical, or other specialized [***487] knowledge will assist the trier of fact to understand the evidence or to determine a fact in issue, . . . an expert . . . may testify thereto . . .". Fed. Rule Evid. 702. It stresses that the subject of the expert's testimony must be "scientific . . . knowledge", and points out that "scientific" "implies a grounding in the methods and procedures of science" and that the word "knowledge" "connotes more than subjective belief or unsupported speculation". Ante, at 590. From this it concludes that "scientific knowledge" must be "derived by the scientific method". Ibid. Proposed testimony, we are told, must be supported by "appropriate validation". Ibid. Indeed, in footnote 9, the Court decides that "in a case involving scientific evidence, evidentiary [*600] reliability will be based upon scientific validity." Ante, at 591, n. 9 (emphasis in original).

Questions arise simply from reading this part of the Court's opinion, and countless more questions will surely arise when hundreds of district judges try to apply its teaching to particular offers of expert testimony. Does all of this dicta apply to an expert seeking to testify on the basis of "technical or other specialized knowledge" -- the other types of expert knowledge to which Rule 702 applies -- or are the "general observations" limited only to "scientific knowledge"? What is the difference between scientific knowledge and technical knowledge; does Rule 702 actually contemplate that the phrase "scientific, technical, or other specialized knowledge" be broken down into numerous subspecies of expertise, or did its authors simply pick general descriptive language covering the sort of expert testimony which courts have customarily received? The Court speaks of its confidence that federal judges can make a "preliminary assessment of whether the reasoning or methodology underlying the testimony is scientifically valid and of whether that reasoning or methodology properly can be applied to the facts in issue." Ante, at 592-593. The Court then states that a "key question" to be answered in deciding whether something is "scientific knowledge" "will be whether it can be (and has been) tested." Ante, at 593. Following this sentence are three quotations from treatises, which not only speak of empirical testing, but one of which states that the "'criterion of the scientific status of a theory is its falsifiability, or refutability, or testability.'" Ibid.

I defer to no one in my confidence in federal judges; but I am at a loss to know what is meant when it is said that the scientific status of a theory depends on its "falsifiability," and I suspect some of them will be, too.

I do not doubt that Rule 702 confides to the judge some gatekeeping responsibility in deciding questions of the admissibility of proffered expert testimony. But I do not think [*601] it imposes on them either the obligation or the authority to become amateur scientists in order to perform that role. I think the Court would be far better advised in this case to decide only the questions presented, and to leave the further development of this important area of the law to future cases.

REFERENCES:

32B Am Jur 2d, Federal Rules of Evidence 435

12 Federal Procedure, L Ed, Evidence 33:131; 33 Federal Procedure, L Ed, Witnesses 80: 133

2 Am Jur Trials 585, Selecting and Preparing Expert Witness; 3 Am Jur Trials 427, Preparing and Using Experimental Evidence

28 USCS Appx, Federal Rules of Evidence, Rule 702

Am Law Prod Liab 3d 54:70, 54:71, 89:38

Ausman & Snyder's Medical Library, L Ed, Pediatrics 5:102

L Ed Digest, Evidence 641
L Ed Index, Experiments or Tests; Expert and Opinion Evidence; Rules of Evidence
ALR Index, Evidence Rules; Experiments and Tests; Expert and Opinion Evidence; Frye Test; Science and Scientific Matters

Annotation References:

Reliability of scientific technique and its acceptance within scientific community as affecting admissibility, at federal trial, of expert testimony as to result of test or study based on such technique--modern cases. 105 ALR Fed 299.
When will expert testimony "assist trier of fact" so as to be admissible at federal trial under Rule 702 of Federal Rules of Evidence. 75 ALR Fed 461.

GENERAL ELECTRIC COMPANY, ET AL., PETITIONERS v. ROBERT K. JOINER ET UX.

NO. 96-188 [3]

SUPREME COURT OF THE UNITED STATES

522 U.S. 136; 118 S. CT. 512; 139 L. ED. 2D 508; 1997 U.S. LEXIS 7503; 66 U.S.L.W. 4036; 48 FED. R. EVID. SERV. (CALLAGHAN) 1; 18 OSHC (BNA) 1097; CCH PROD. LIAB. REP. P15,120; 97 CAL. DAILY OP. SERVICE 9355; 97 DAILY JOURNAL DAR 15051; 28 ELR 20227; 1997 COLO. J. C.A.R. 3361; 11 FLA. L. WEEKLY FED. S 284

OCTOBER 14, 1997, ARGUED
DECEMBER 15, 1997, DECIDED

PRIOR HISTORY: ON WRIT OF CERTIORARI TO THE UNITED STATES COURT OF APPEALS FOR THE ELEVENTH CIRCUIT.

DISPOSITION: 78 F.3d 524, reversed and remanded.

DECISION:
Abuse of discretion held to be proper standard for review of Federal District Court's decision to admit or exclude expert scientific testimony, and court held not to have abused discretion in excluding such testimony.

SUMMARY:
An electrician, who alleged that his small cell lung cancer was promoted by on-the-job exposure to polychlorinated biphenyls (PCBs) and to furans and dioxins (some PCB derivatives), sued in a Georgia state court the manufacturers of the products through which the exposure had occurred. After the manufacturers removed the case to the United States District Court for the Northern District of Georgia, the District Court--in excluding the proffered scientific testimony of the electrician's experts indicating a link between exposure to PCBs and small cell lung cancer, and in granting the manufacturers' summary judgment motion--expressed the view that (1) there was no genuine issue as to whether the electrician had been

exposed to furans and dioxins, and (2) the expert testimony did not rise above subjective belief or unsupported speculation (864 F. Supp 1310). The United States Court of Appeals for the Eleventh Circuit, in reversing the District Court's judgment, expressed the view that (1) the Court of Appeals would apply a particularly stringent standard of review to a trial judge's exclusion of expert testimony, and (2) under that standard, the District Court had erred in excluding the testimony of the electrician's experts (78 F.3d 524).

On certiorari, the United States Supreme Court reversed and remanded. In an opinion by Rehnquist, Ch. J., expressing the unanimous view of the court as to holding 1 below, and joined by O'Connor, Scalia, Kennedy, Souter, Thomas, Ginsburg, and Breyer, JJ,, as to holding 2 below, it was held that (1) abuse of discretion is the proper standard for an appellate court to apply in reviewing a Federal District Court's decision to admit or exclude expert scientific testimony at trial; and (2) because it was within the discretion of the District Court in the instant case to conclude that the animal studies and the four epidemiological studies upon which the experts relied were not sufficient, whether individually or in combination, to support the experts' conclusions that the electrician's exposure to PCBs contributed to his cancer, the District Court did not abuse its discretion in excluding the experts' testimony.

Breyer, J., concurring, expressed the view that given the offer of cooperative effort from the scientific to the legal community, and given the various methods authorized under the Federal Rules of Evidence and the Federal Rules of Civil Procedure for facilitating the trial courts' task as the gatekeepers insuring that scientific testimony or evidence admitted at trial is relevant and reliable, the gatekeeping requirement would not prove inordinately difficult to implement.

Stevens, J., concurring in part and dissenting in part, (1) agreed that abuse of discretion was the proper standard for an appellate court's review of a Federal District Court's admission or exclusion of expert scientific evidence at trial, but (2) as to the question whether the District Court in the instant case had properly held the expert testimony to be inadmissible, expressed the view that (a) it was not certain that the parties had adequately briefed the question or that the Supreme Court had adequately explained why the Court of Appeals' disposition was erroneous, and (b) the case ought to have been remanded to the Court of Appeals for application of the proper standard of review.

LAWYERS' EDITION HEADNOTES:

[***LEdHN1]
APPEAL §1391
expert scientific testimony -- admission or exclusion -- abuse of discretion --
Headnote: [1A] [1B] [1C]

Abuse of discretion is the proper standard for an appellate court to apply in reviewing a Federal District Court's decision to admit or exclude expert scientific testimony at trial, because all evidentiary decisions are reviewed under an abuse-of-discretion standard, where the United States Supreme Court has held that abuse of discretion is the proper standard of review of a District

Court's evidentiary rulings.

[***LEdHN2]
EVIDENCE §643
expert scientific testimony -- linkage -- exclusion --
Headnote: [2A] [2B] [2C] [2D]

A Federal District Court, in entering summary judgment for the manufacturers who have been sued by an electrician who alleges that the electrician's small cell lung cancer was promoted by on-the-job exposure to the manufacturer's products that contained polychlorinated biphenyls (PCBs), does not abuse the court's discretion by excluding at trial the testimony of the electrician's scientific experts indicating a link between PCB exposure and small cell lung cancer, because it is within the court's discretion to conclude that the animal studies and the four epidemiological studies on which the experts rely are not sufficient, whether individually or in combination, to support the experts' conclusions that the electrician's exposure to PCBs contributed to the electrician's cancer, where (1) the animal studies, which involved the injection of massive doses of highly concentrated PCBs directly into the peritoneums or stomachs of infant mice, were so dissimilar to the facts presented in the instant litigation; and (2) as to the epidemiological studies, (a) the authors of the first study, which involved capacitor plant workers who had been exposed to PCBs, were unwilling to say that PCB exposure had caused cancer among the workers the authors examined, (b) the increase in lung cancer among the PCB production plant workers involved in the second study was not statistically significant, and the authors of the study did not suggest a link between the increase in lung cancer deaths and the exposure to PCBs, (c) the third study, which involved cable manufacturing workers who had been exposed to mineral oil, made no mention of PCBs and was expressly limited to the type of mineral oil involved in that study, and (d) although the fourth study involved a PCB-exposed group, the subjects of that study had been exposed to numerous potential carcinogens, including toxic rice oil that the subjects had ingested. (Stevens, J., dissented from this holding.)

[***LEdHN3]
APPEAL §1391
EVIDENCE §641
scientific testimony -- admission or exclusion -- discretion --
Headnote: [3]

With respect to the admission of scientific testimony in Federal District Courts, the Federal Rules of Evidence leave in place the "gatekeeper" role of the trial judge in screening such evidence; a Federal Court of Appeals applying abuse-of-discretion review to such rulings may not categorically distinguish between rulings allowing expert testimony and rulings which disallow it.

[***LEdHN4]
APPEAL §1392
SUMMARY JUDGMENT AND JUDGMENT ON PLEADINGS §3

resolving motion -- exclusion of expert testimony -- abuse of discretion --
Headnote: [4]

On certiorari to review a Federal Court of Appeals' judgment--which, in reversing a Federal District Court's granting of a summary judgment motion by the manufacturers who were sued by an electrician who alleged that the electrician's small cell lung cancer was promoted by on-the-job exposure to the manufacturer's products that contained polychlorinated biphenyls (PCBs), held that the District Court had erred in concluding that the testimony of the electrician's experts indicating a link between PCB exposure and small cell lung cancer was inadmissible for failure to rise above subjective belief or unsupported speculation--although on a motion for summary judgment, disputed issues of fact are resolved against the moving party, the question of admissibility of expert testimony is not such an issue of fact; thus, the United States Supreme Court will reject the electrician's argument that because the granting of summary judgment for the manufacturers was outcome determinative, it should have been subjected to a more searching standard of appellate review than the abuse-of-discretion standard that the Supreme Court holds to be the proper standard.

[***LEdHN5]
APPEAL §1392
exclusion of expert testimony -- abuse of discretion --
Headnote: [5]

In a suit in which a cancer victim seeks to link the victim's development of cancer to the victim's exposure to polychlorinated biphenyls (PCBs) and their derivatives, a Federal Court of Appeals--which states that it will apply a particularly stringent standard of review to the trial judge's exclusion of expert testimony--errs in the Court of Appeals' review of a Federal District Court's exclusion, at trial, of the victim's expert testimony, as, in applying an overly stringent review to the trial court's ruling, the Court of Appeals fails to give the trial court the deference that is the hallmark of abuse-of-discretion review.

[***LEdHN6]
EVIDENCE §641
expert opinion -- admission at trial --
Headnote: [6]

With respect to the issue of admission of expert testimony at trial, although trained experts commonly extrapolate from existing data, nothing in either a United States Supreme Court decision--indicating that the focus must be solely on principles and methodology, not on the conclusions that they generate--or the Federal Rules of Evidence requires a Federal District Court to admit opinion evidence which is connected to existing data by only the expert's own statement, as a court may conclude that there is simply too great an analytical gap between the data and the opinion proffered. (Stevens, J., dissented from this holding.)

[***LEdHN7]
APPEAL §1681

reversal -- remand -- leaving question open --
Headnote: [7]

On certiorari to determine what standard a federal appellate court should apply in reviewing a federal trial court's decision to admit or exclude expert testimony, the United States Supreme Court, in holding that abuse of discretion is the appropriate standard and that the Federal District Court involved in the instant case did not abuse its discretion when it excluded certain proffered expert testimony that indicated a link between exposure to polychlorinated biphenyls (PCBs) and small cell lung cancer, will reverse a Federal Court of Appeals judgment--which, in reversing a summary judgment for some manufacturers who were sued by an electrician who alleged that the electrician's small cell lung cancer was promoted by on-the-job exposure to PCBs and to furans and dioxins (some PCB derivatives), held that the District Court had erred in concluding that (1) there was no genuine issue of material fact as to whether the electrician had been exposed to furans and dioxins, and (2) the linkage testimony of the electrician's experts was inadmissible for failure to rise above subjective belief or unsupported speculation--and will remand the case for proceedings consistent with the Supreme Court's opinion, where the manufacturers, in their petition to the Supreme Court, have not challenged the Court of Appeals' reversal of the District Court's determination that there was no genuine issue as to whether the electrician had been exposed to furans and dioxins; thus, whether the electrician was exposed to furans and dioxins and whether if there was such exposure, the opinions of the electrician's experts would then be admissible remain open questions.

SYLLABUS: After he was diagnosed with small-cell lung cancer, respondent Joiner sued in Georgia state court, alleging, inter alia, that his disease was "promoted" by his workplace exposure to chemical "PCBs" and derivative "furans" and "dioxins" that were manufactured by, or present in materials manufactured by, petitioners. Petitioners removed the case to federal court and moved for summary judgment. Joiner responded with the depositions of expert witnesses, who testified that PCBs, furans, and dioxins can promote cancer, and opined that Joiner' exposure to those chemicals was likely responsible for his cancer. The District Court ruled that there was a genuine issue of material fact as to whether Joiner had been exposed to PCBs, but granted summary judgment for petitioners because (1) there was no genuine issue as to whether he had been exposed to furans and dioxins, and (2) his experts' testimony had failed to show that there was a link between exposure to PCBs and small-cell lung cancer and was therefore inadmissible because it did not rise above "subjective belief or unsupported speculation." In reversing, the Eleventh Circuit applied "a particularly stringent standard of review" to hold that the District Court had erred in excluding the expert testimony.

Held:

1. Abuse of discretion -- the standard ordinarily applicable to review of evidentiary rulings -- is the proper standard by which to review a district court's decision to admit or exclude expert scientific evidence. Contrary to the Eleventh Circuit's suggestion, Daubert v. Merrell Dow Pharmaceuticals, Inc., 509 U.S. 579, 125 L. Ed. 2d 469, 113 S. Ct. 2786, did not somehow alter this general rule in the

context of a district court's decision to exclude scientific evidence. Daubert did not address the appellate review standard for evidentiary rulings at all, but did indicate that, while the Federal Rules of Evidence allow district courts to admit a somewhat broader range of scientific testimony than did pre-existing law, they leave in place the trial judge's "gatekeeper" role of screening such evidence to ensure that it is not only relevant, but reliable. Id., at 589. A court of appeals applying "abuse of discretion" review to such rulings may not categorically distinguish between rulings allowing expert testimony and rulings which disallow it. Compare Beech Aircraft Corp. v. Rainey, 488 U.S. 153, 172, 102 L. Ed. 2d 445, 109 S. Ct. 439, with United States v. Abel, 469 U.S. 45, 54, 83 L. Ed. 2d 450, 105 S. Ct. 465. This Court rejects Joiner's argument that because the granting of summary judgment in this case was "outcome determinative," it should have been subjected to a more searching standard of review. On a summary judgment motion, disputed issues of fact are resolved against the moving party -- here, petitioners. But the question of admissibility of expert testimony is not such an issue of fact, and is reviewable under the abuse of discretion standard. In applying an overly "stringent" standard, the Eleventh Circuit failed to give the trial court the deference that is the hallmark of abuse of discretion review. Pp. 4-5.

2. A proper application of the correct standard of review indicates that the District Court did not err in excluding the expert testimony at issue. The animal studies cited by respondent's experts were so dissimilar to the facts presented here -- i.e., the studies involved infant mice that developed alveologenic adenomas after highly concentrated, massive doses of PCBs were injected directly into their peritoneums or stomachs, whereas Joiner was an adult human whose small-cell carcinomas allegedly resulted from exposure on a much smaller scale -- that it was not an abuse of discretion for the District Court to have rejected the experts' reliance on those studies. Nor did the court abuse its discretion in concluding that the four epidemiological studies on which Joiner relied were not a sufficient basis for the experts' opinions, since the authors of two of those studies ultimately were unwilling to suggest a link between increases in lung cancer and PCB exposure among the workers they examined, the third study involved exposure to a particular type of mineral oil not necessarily relevant here, and the fourth involved exposure to numerous potential carcinogens in addition to PCBs. Nothing in either Daubert or the Federal Rules of Evidence requires a district court to admit opinion evidence which is connected to existing data only by the ipse dixit of the expert. Pp. 6-9.

3. These conclusions, however, do not dispose of the entire case. The Eleventh Circuit reversed the District Court's conclusion that Joiner had not been exposed to furans and dioxins. Because petitioners did not challenge that determination in their certiorari petition, the question whether exposure to furans and dioxins contributed to Joiner's cancer is still open. Pp. 9-10.

78 F.3d 524, reversed and remanded.

COUNSEL: Steven R. Kuney argued the cause for petitioners. With him on the briefs were John G. Kester, David H. Flint, Alexander J. Simmons, Jr., Henry W. Ewalt, and Gerard H. Davidson, Jr.

Deputy Solicitor General Wallace argued the cause for the United States as amicus curiae urging reversal. With him on the brief were Acting Solicitor General

Dellinger, Assistant Attorney General Hunger, Edward C. DuMont, and John P. Schnitker.

Michael H. Gottesman argued the cause for respondents. With him on the brief were Kenneth J. Chesebro, David L. Shapiro, and Michael J. Warshauer.

Briefs of amici curiae urging reversal were filed for the Chamber of Commerce of the United States by Thomas S. Martin, Stephen A. Bokat, and Robin S. Conrad; for the American Medical Association by Jack R. Bierig, Carter G. Phillips, Kirk B. Johnson, and Michael L. Ile; for the Chemical Manufacturers Association by Bert Black, David J. Schenck, and Donald D. Evans; for the Dow Chemical Company by John E. Muench and Robert M. Dow, Jr.; for the Pharmaceutical Research and Manufacturers of America by Bruce N. Kuhlik; for the Washington Legal Foundation by Arvin Maskin, Gerald A. Stein, Daniel J. Popeo, and Paul D. Kamenar; and for Bruce Ames et al. by Martin S. Kaufman and Douglas Foster.

Briefs of amici curiae urging affirmance were filed for the Trial Lawyers for Public Justice by Steven E. Fineman and Arthur H. Bryant; for the Association of Trial Lawyers of America by Jeffrey Robert White; for Ardith Cavallo by William A. Beeton, Jr.; and for Peter Orris, M.D., et al. by Gerson H. Smoger.

Briefs of amici curiae were filed for the New England Journal of Medicine et al. by Margaret S. Woodruff and Arlin M. Adams; and for the Product Liability Advisory Council, Inc., et al. by Mary A. Wells, Jan S. Amundson, and Quentin Riegel.

JUDGES: REHNQUIST, C. J., delivered the opinion for a unanimous Court with respect to Parts I and II, and the opinion of the Court with respect to Part III, in which O'CONNOR, SCALIA, KENNEDY, SOUTER, THOMAS, GINSBURG, and BREYER, JJ., joined. BREYER, J., filed a concurring opinion. STEVENS, J., filed an opinion concurring in part and dissenting in part.

OPINIONBY: REHNQUIST

OPINION: [**515] [***514] [*138] CHIEF JUSTICE REHNQUIST delivered the opinion of the Court.

[***LEdHR1A] [1A] [***LEdHR2A] [2A]We granted certiorari in this case to determine what standard an appellate court should apply in reviewing a trial [*139] court's decision to admit or exclude expert testimony under Daubert v. Merrell Dow Pharmaceuticals, Inc., 509 U.S. 579, 125 L. Ed. 2d 469, 113 S. Ct. 2786 (1993). We hold that abuse of discretion is the appropriate standard. We apply this standard and conclude that the District Court in this case did not abuse its discretion when it excluded certain proffered expert testimony.

I

Respondent Robert Joiner began work as an electrician in the Water & Light Department of Thomasville, Georgia (City) in 1973. This job required him to work with and around the City's electrical transformers, which used a mineral-based dielectric fluid [**516] as a coolant. Joiner often had to stick his hands and arms into the fluid to make repairs. The fluid would sometimes splash onto him, occasionally getting into his eyes and mouth. In 1983 the City discovered that the fluid in some of the transformers was contaminated with polychlorinated biphenyls (PCBs). PCBs are widely considered to be hazardous to human health. Congress, with limited exceptions, banned the production and sale of PCBs in 1978. See 90 Stat. 2020, 15 U.S.C. § 2605(e)(2)(A).

Joiner was diagnosed with small cell lung cancer in 1991. He n1 sued petitioners in Georgia state court the following year. Petitioner Monsanto manufactured PCBs from 1935 to 1977; petitioners General Electric and Westinghouse Electric manufactured transformers and dielectric fluid. In his complaint Joiner linked his development of cancer to his exposure to PCBs and their derivatives, polychlorinated dibenzofurans (furans) and polychlorinated dibenzodioxins (dioxins). Joiner had been a smoker for approximately eight years, his parents had both been smokers, and there was a history of lung cancer in his family. He was thus perhaps already at a heightened risk of developing lung cancer eventually. The suit alleged that his exposure to PCBs "promoted" [*140] his cancer; [***515] had it not been for his exposure to these substances, his cancer would not have developed for many years, if at all.

- - - - - - - - - - - - - - Footnotes - - - - - - - - - - - - - -
n1 Joiner's wife was also a plaintiff in the suit and is a respondent here. For convenience, we refer to respondent in the singular.
- - - - - - - - - - - - End Footnotes- - - - - - - - - - - - -

Petitioners removed the case to federal court. Once there, they moved for summary judgment. They contended that (1) there was no evidence that Joiner suffered significant exposure to PCBs, furans, or dioxins, and (2) there was no admissible scientific evidence that PCBs promoted Joiner's cancer. Joiner responded that there were numerous disputed factual issues that required resolution by a jury. He relied largely on the testimony of expert witnesses. In depositions, his experts had testified that PCBs alone can promote cancer and that furans and dioxins can also promote cancer. They opined that since Joiner had been exposed to PCBs, furans, and dioxins, such exposure was likely responsible for Joiner's cancer.

The District Court ruled that there was a genuine issue of material fact as to whether Joiner had been exposed to PCBs. But it nevertheless granted summary judgment for petitioners because (1) there was no genuine issue as to whether Joiner had been exposed to furans and dioxins, and (2) the testimony of Joiner's experts had failed to show that there was a link between exposure to PCBs and small cell lung cancer. The court believed that the testimony of respondent's experts to the contrary did not rise above "subjective belief or unsupported speculation." 864 F. Supp. 1310, 1326 (ND Ga. 1994). Their testimony was therefore inadmissible.

The Court of Appeals for the Eleventh Circuit reversed. 78 F.3d 524 (1996). It held that "because the Federal Rules of Evidence governing expert testimony display a preference for admissibility, we apply a particularly stringent standard of review to the trial judge's exclusion of expert testimony." Id. at 529. Applying that standard, the Court of Appeals held that the District Court had erred in excluding the testimony of Joiner's expert witnesses. The [*141] District Court had made two fundamental errors. First, it excluded the experts' testimony because it "drew different conclusions from the research than did each of the experts." The Court of Appeals opined that a district court should limit its role to determining the "legal reliability of proffered expert testimony, leaving the jury to decide the correctness of competing expert opinions." Id. at 533. Second, the District Court had held that there was no genuine issue of material fact as to whether Joiner had been exposed to furans and dioxins. This was also incorrect, said the Court of Appeals, because testimony in the record supported the proposition that there had been such exposure.

We granted petitioners' petition for a writ of certiorari, 520 U.S. (1997), and we now reverse.

[**517] II

[***LEdHR1B] [1B]Petitioners challenge the standard applied by the Court of Appeals in reviewing the District Court's decision to exclude respondent's experts' proffered testimony. They argue that that court should have applied traditional "abuse of discretion" review. Respondent agrees that abuse of discretion is the correct standard of review. He contends, however, that the Court of Appeals applied an abuse of discretion standard [***516] in this case. As he reads it, the phrase "particularly stringent" announced no new standard of review. It was simply an acknowledgement that an appellate court can and will devote more resources to analyzing district court decisions that are dispositive of the entire litigation. All evidentiary decisions are reviewed under an abuse of discretion standard. He argues, however, that it is perfectly reasonable for appellate courts to give particular attention to those decisions that are outcome-determinative.

We have held that abuse of discretion is the proper standard of review of a district court's evidentiary rulings. Old Chief v. United States, 519 U.S. , n. 1, 117 S. Ct. 644, 136 L. Ed. 2d 574 (1997) (slip op., at 1-2, n.1), United States v. Abel, 469 U.S. 45, 54, 83 L. Ed. 2d 450, 105 S. Ct. 465 (1984). Indeed, our cases on [*142] the subject go back as far as Spring Co. v. Edgar, 99 U.S. 645, 658, 25 L. Ed. 487 (1879) where we said that "cases arise where it is very much a matter of discretion with the court whether to receive or exclude the evidence; but the appellate court will not reverse in such a case, unless the ruling is manifestly erroneous." The Court of Appeals suggested that Daubert somehow altered this general rule in the context of a district court's decision to exclude scientific evidence. But Daubert did not address the standard of appellate review for evidentiary rulings at all. It did hold that the "austere" Frye standard of "general acceptance" had not been carried over into the Federal Rules of Evidence. But the opinion also said:

> "That the Frye test was displaced by the Rules of Evidence does not mean, however, that the Rules themselves place no limits on the admissibility of purportedly scientific evidence. Nor is the trial judge disabled from screening such evidence. To the contrary, under the Rules the trial judge must ensure that any and all scientific testimony or evidence admitted is not only relevant, but reliable." 509 U.S. at 589 (footnote omitted).

[***LEdHR3] [3] [***LEdHR4] [4]Thus, while the Federal Rules of Evidence allow district courts to admit a somewhat broader range of scientific testimony than would have been admissible under Frye, they leave in place the "gatekeeper" role of the trial judge in screening such evidence. A court of appeals applying "abuse of discretion" review to such rulings may not categorically distinguish between rulings allowing expert testimony and rulings which disallow it. Compare Beech Aircraft Corp v. Rainey, 488 U.S. 153, 172, 102 L. Ed. 2d 445, 109 S. Ct. 439 (1988) (applying abuse of discretion review to a lower court's decision to exclude evidence) with United States v. Abel, supra at 54 (applying abuse of discretion review to a lower court's decision to admit evidence). We likewise reject respondent's argument that because the granting of summary judgment in this case [*143] was "outcome determinative," it should have been subjected to a more searching standard of review. On a motion for summary judgment, disputed issues of fact are resolved against the moving party -- here, petitioners. But the question of admissibility of expert testimony is not such an issue of fact, and is reviewable under the abuse of discretion standard. [***517]

[***LEdHR5] [5]We hold that the Court of Appeals erred in its review of the exclusion of Joiner's experts' testimony. In applying an overly "stringent" review to that ruling, it failed to give the trial court the deference that is the hallmark of abuse of discretion review. See, e.g., Koon v. United States, 518 U.S. 81, , 135 L. Ed. 2d 392, 116 S. Ct. 2035 (1996)(slip op., at 14-15).

III

[***LEdHR2B] [2B]We believe that a proper application of the correct standard of review here indicates that the District Court did not abuse its [**518] discretion. Joiner's theory of liability was that his exposure to PCBs and their derivatives "promoted" his development of small cell lung cancer. In support of that theory he proffered the deposition testimony of expert witnesses. Dr. Arnold Schecter testified that he believed it "more likely than not that Mr. Joiner's lung cancer was causally linked to cigarette smoking and PCB exposure." App. at 107. Dr. Daniel Teitelbaum testified that Joiner's "lung cancer was caused by or contributed to in a significant degree by the materials with which he worked." Id. at 140.

Petitioners contended that the statements of Joiner's experts regarding causation were nothing more than speculation. Petitioners criticized the testimony of the experts in that it was "not supported by epidemiological studies . . . [and was] based exclusively on isolated studies of laboratory animals." Joiner responded by claiming that his experts had identified "relevant animal studies which support their opinions." [*144] He also directed the court's attention to four epidemiological studies n2 on which his experts had relied.

- - - - - - - - - - - - - - Footnotes - - - - - - - - - - - - - -

n2 Epidemiological studies examine the pattern of disease in human populations.

- - - - - - - - - - - - End Footnotes- - - - - - - - - - - - - -

The District Court agreed with petitioners that the animal studies on which respondent's experts relied did not support his contention that exposure to PCBs had contributed to his cancer. The studies involved infant mice that had developed cancer after being exposed to PCBs. The infant mice in the studies had had massive doses of PCBs injected directly into their peritoneums n3 or stomachs. Joiner was an adult human being whose alleged exposure to PCBs was far less than the exposure in the animal studies. The PCBs were injected into the mice in a highly concentrated form. The fluid with which Joiner had come into contact generally had a much smaller PCB concentration of between 0-500 parts per million. The cancer that these mice developed was alveologenic adenomas; Joiner had developed small-cell carcinomas. No study demonstrated that adult mice developed cancer after being exposed to PCBs. One of the experts admitted that no study had demonstrated that PCBs lead to cancer in any other species.

- - - - - - - - - - - - - - Footnotes - - - - - - - - - - - - - - -
n3 The peritoneum is the lining of the abdominal cavity.
- - - - - - - - - - - - End Footnotes- - - - - - - - - - - - - -

Respondent failed to reply to this criticism. Rather than explaining how and why the experts could have extrapolated their opinions from these seemingly far-removed animal studies, respondent chose "to proceed as if the only issue [was] whether animal studies can ever be a proper foundation for an expert's opinion." Joiner, 864 F. Supp. at 1324. Of course, whether animal studies can ever be a proper foundation for an expert's opinion was not the issue. [***518] The issue was whether these experts' opinions were sufficiently supported by the animal studies on which they purported to rely. The studies were so dissimilar to the facts presented in this litigation [*145] that it was not an abuse of discretion for the District Court to have rejected the experts' reliance on them.

The District Court also concluded that the four epidemiological studies on which respondent relied were not a sufficient basis for the experts' opinions. The first such study involved workers at an Italian capacitor n4 plant who had been exposed to PCBs. Bertazzi, Riboldi, Pesatori, Radice, & Zocchetti, Cancer Mortality of Capacitor Manufacturing Workers, 11 American Journal of Industrial Medicine 165 (1987). The authors noted that lung cancer deaths among ex-employees at the plant were higher than might have been expected, but concluded that "there were apparently no grounds for associating lung cancer deaths (although increased above expectations) and exposure in the plant." Id. at 172. Given that Bertazzi et al. were unwilling to say that PCB exposure had caused cancer among the workers they examined, their study did not support the experts' conclusion that Joiner's exposure to PCBs caused his cancer.

- - - - - - - - - - - - - - Footnotes - - - - - - - - - - - - - - -
n4 A capacitor is an electrical component that stores an electric charge.

- - - - - - - - - - - - End Footnotes- - - - - - - - - - - - - -

The second study followed employees who had worked at Monsanto's PCB production plant. J. Zack & D. Munsch, Mortality [**519] of PCB Workers at the Monsanto Plant in Sauget, Illinois (Dec. 14, 1979)(unpublished report), 3 Rec., Doc. No. 11. The authors of this study found that the incidence of lung cancer deaths among these workers was somewhat higher than would ordinarily be expected. The increase, however, was not statistically significant and the authors of the study did not suggest a link between the increase in lung cancer deaths and the exposure to PCBs.

The third and fourth studies were likewise of no help. The third involved workers at a Norwegian cable manufacturing company who had been exposed to mineral oil. Ronneberg, Andersen, Skyberg, Mortality and Incidence of Cancer Among Oil-Exposed Workers in a Norwegian Cable Manufacturing Company, 45 British Journal of Industrial [*146] Medicine 595 (1988). A statistically significant increase in lung cancer deaths had been observed in these workers. The study, however, (1) made no mention of PCBs and (2) was expressly limited to the type of mineral oil involved in that study, and thus did not support these experts' opinions. The fourth and final study involved a PCB-exposed group in Japan that had seen a statistically significant increase in lung cancer deaths. Kuratsune, Nakamura, Ikeda, & Hirohata, Analysis of Deaths Seen Among Patients with Yusho -- A Preliminary Report, 16 Chemosphere, Nos. 8/9, 2085 (1987). The subjects of this study, however, had been exposed to numerous potential carcinogens, including toxic rice oil that they had ingested.

[***LEdHR2C] [2C] [***LEdHR6] [6]Respondent points to Daubert's language that the "focus, of course, must be solely on principles and methodology, not on the conclusions that they generate." 509 U.S. at 595. He claims that because the District Court's disagreement was with the [***519] conclusion that the experts drew from the studies, the District Court committed legal error and was properly reversed by the Court of Appeals. But conclusions and methodology are not entirely distinct from one another. Trained experts commonly extrapolate from existing data. But nothing in either Daubert or the Federal Rules of Evidence requires a district court to admit opinion evidence which is connected to existing data only by the ipse dixit of the expert. A court may conclude that there is simply too great an analytical gap between the data and the opinion proffered. See Turpin v. Merrell Dow Pharmaceuticals, Inc., 959 F.2d 1349, 1360 (CA 6), cert. denied, 506 U.S. 826, 121 L. Ed. 2d 47, 113 S. Ct. 84 (1992). That is what the District Court did here, and we hold that it did not abuse its discretion in so doing.

[***LEdHR1C] [1C] [***LEdHR2D] [2D]We hold, therefore, that abuse of discretion is the proper standard by which to review a district court's decision to admit or exclude scientific evidence. We further hold that, because it was within the District Court's discretion to conclude that the studies upon which the experts relied were not [*147] sufficient, whether individually or in combination, to support their conclusions that Joiner's exposure to PCBs contributed to his cancer, the District Court did not abuse its discretion in excluding their testimony. These conclusions, however, do not dispose of this entire case.

[***LEdHR7] [7]Respondent's original contention was that his exposure to PCBs, furans, and dioxins contributed to his cancer. The District Court ruled that there was a genuine issue of material fact as to whether Joiner had been exposed to PCBs, but concluded that there was no genuine issue as to whether he had been exposed to furans and dioxins. The District Court accordingly never explicitly considered if there was admissible evidence on the question whether Joiner's alleged exposure to furans and dioxins contributed to his cancer. The Court of Appeals reversed the District Court's conclusion that there had been no exposure to furans and dioxins. Petitioners did not challenge this determination in their petition to this Court. Whether Joiner was exposed to furans and dioxins, and whether if there was such exposure, the opinions of Joiner's experts would then be admissible, remain open questions. We accordingly reverse the judgment of the Court of Appeals and remand this case for proceedings consistent with this opinion.

It is so ordered.

CONCURBY: BREYER; STEVENS (In Part)
CONCUR: [**520]
JUSTICE BREYER, concurring.

The Court's opinion, which I join, emphasizes Daubert's statement that a trial judge, acting as "gatekeeper," must "'ensure that any and all scientific testimony or evidence admitted is not only relevant, but reliable.'" Ante, at 5 (quoting Daubert v. Merrell Dow Pharmaceuticals, Inc., 509 U.S. 579, 589, 125 L. Ed. 2d 469, 113 S. Ct. 2786 (1993)). This requirement will sometimes ask judges to make subtle and sophisticated determinations about scientific methodology and its relation to the conclusions an expert witness seeks to offer – particularly when a case arises in an area where [***520] the science itself is tentative or [*148] uncertain, or where testimony about general risk levels in human beings or animals is offered to prove individual causation. Yet, as amici have pointed out, judges are not scientists and do not have the scientific training that can facilitate the making of such decisions. See, e.g., Brief for Trial Lawyers for Public Justice as Amicus Curiae 15; Brief for The New England Journal of Medicine et al. as Amici Curiae 2 ("Judges . . . are generally not trained scientists ").

Of course, neither the difficulty of the task nor any comparative lack of expertise can excuse the judge from exercising the "gatekeeper" duties that the Federal Rules impose -- determining, for example, whether particular expert testimony is reliable and "will assist the trier of fact," Fed. Rule Evid. 702, or whether the "probative value" of testimony is substantially outweighed by risks of prejudice, confusion or waste of time. Fed. Rule Evid. 403. To the contrary, when law and science intersect, those duties often must be exercised with special care.

Today's toxic tort case provides an example. The plaintiff in today's case says that a chemical substance caused, or promoted, his lung cancer. His concern, and that of others, about the causes of cancer is understandable, for cancer kills over one in five Americans. See U.S. Dept. of Health and Human Services, National Center for Health Statistics, Health United States 1996-97 and Injury Chartbook 117 (1997) (23.3% of all deaths in 1995). Moreover, scientific evidence implicates some chemicals as potential causes of some cancers. See, e.g., U.S. Dept. of Health and Human Services, Public Health Service, National Toxicology Program, 1 Seventh Annual Report on Carcinogens, pp. v-vi (1994). Yet modern life, including good health as well as economic well-being, depends upon the use of artificial or manufactured substances, such as chemicals. And it may, therefore, prove particularly important to see that judges fulfill their Daubert gatekeeping function, so that they help assure that the powerful engine of tort liability, which can generate [*149] strong financial incentives to reduce, or to eliminate, production, points towards the right substances and does not destroy the wrong ones. It is, thus, essential in this science-related area that the courts administer the Federal Rules of Evidence in order to achieve the "ends" that the Rules themselves set forth, not only so that proceedings may be "justly determined," but also so "that the truth may be ascertained." Fed. Rule Evid. 102.

I therefore want specially to note that, as cases presenting significant science-related issues have increased in number, see Judicial Conference of the United States, Report of the Federal Courts Study Committee 97 (Apr. 2, 1990) ("Economic, statistical, technological, and natural and social scientific data are becoming increasingly important in both routine and complex litigation"), judges have increasingly found in the Rules of Evidence and Civil Procedure ways to help them overcome the inherent difficulty of making determinations about complicated scientific or otherwise technical evidence. Among these techniques are an increased use of Rule 16's pretrial conference authority to narrow the scientific issues in dispute, pretrial hearings where potential experts are subject [***521] to examination by the court, and the appointment of special masters and specially trained law clerks. See J. Cecil & T. Willging, Court-Appointed Experts: Defining the Role of Experts Appointed Under Federal Rule of Evidence 706, pp. 83-88 (1993); J. Weinstein, Individual Justice in Mass Tort Litigation 107-110 (1995); cf. Kaysen, In Memoriam: Charles E. Wyzanski, Jr., 100 Harv. L. Rev. 713, 713-715 (1987) (discussing a judge's use of an economist as a law clerk in United States v. United Shoe Machinery Corp., 110 F. Supp. 295 [**521] (D Mass 1953), aff'd, 347 U.S. 521, 98 L. Ed. 910, 74 S. Ct. 699 (1954)).

In the present case, the New England Journal of Medicine has filed an amici brief "in support of neither petitioners nor respondents" in which the Journal writes:

> "[A] judge could better fulfill this gatekeeper function if he or she had help from scientists. Judges should be [*150] strongly encouraged to make greater use of their inherent authority . . . to appoint experts . . . . Reputable experts could be recommended to courts by established scientific organizations, such as the National Academy of Sciences or the American Association for the Advancement of Science."

Brief for The New England Journal of Medicine 18-19; cf. Fed. Rule Evid. 706 (court may "on its own motion or on the motion of any party" appoint an expert to serve on behalf of the court, and this expert may be selected as "agreed upon by the parties" or chosen by the court); see also Weinstein, supra, at 116 (a court should sometimes "go beyond the experts proffered by the parties" and "utilize its powers to appoint independent experts under Rule 706 of the Federal Rules of Evidence"). Given this kind of offer of cooperative effort, from the scientific to the legal community, and given the various Rules-authorized methods for facilitating the courts' task, it seems to me that Daubert's gatekeeping requirement will not prove inordinately difficult to implement; and that it will help secure the basic objectives of the Federal Rules of Evidence; which are, to repeat, the ascertainment of truth and the just determination of proceedings. Fed. Rule Evid. 102.

DISSENTBY: STEVENS (In Part)
DISSENT:
JUSTICE STEVENS, concurring in part and dissenting in part.

The question that we granted certiorari to decide is whether the Court of Appeals applied the correct standard of review. That question is fully answered in Parts I and II of the Court's opinion. Part III answers the quite different question whether the District Court properly held that the testimony of plaintiff 's expert witnesses was inadmissible. Because I am not sure that the parties have adequately briefed that question, or that the Court has adequately explained why the Court of Appeals' disposition was erroneous, I do not join Part III. Moreover, because a proper answer to that question requires a study of the record that can be [*151] performed more efficiently by the Court of Appeals than by the nine members of this Court, I would remand the case to that court for application of the proper standard of review.

One aspect of the record will illustrate my concern. As the Court of Appeals pointed out, Joiner's experts relied on "the studies of at least [***522] thirteen different researchers, and referred to several reports of the World Health Organization that address the question of whether PCBs cause cancer." 78 F.3d 524, 533 (CA11 1996). Only one of those studies is in the record, and only six of them were discussed in the District Court opinion. Whether a fair appraisal of either the methodology or the conclusions of Joiner's experts can be made on the basis of such an incomplete record is a question that I do not feel prepared to answer.

It does seem clear, however, that the Court has not adequately explained why its holding is consistent with Federal Rule of Evidence 702, n1 as interpreted in Daubert v. Merrell Dow Pharmaceuticals, Inc., 509 U.S. 579, 125 L. Ed. 2d 469, 113 S. Ct. 2786 (1993). n2 In general, scientific testimony that is both relevant and reliable must be admitted and testimony that is irrelevant or unreliable [**522] must be excluded. Id., at 597. In this case, the District Court relied on both grounds for exclusion.

- - - - - - - - - - - - - - Footnotes - - - - - - - - - - - - - - -

n1 Rule 702 states: "If scientific, technical, or other specialized knowledge will assist the trier of fact to understand the evidence or to determine a fact in issue, a witness qualified as an expert by knowledge, skill, experience, training, or education, may testify thereto in the form of an opinion or otherwise."

n2 The specific question on which the Court granted certiorari in Daubert was whether the rule of Frye v. United States, 54 App. D.C. 46, 293 F. 1013 (1923),

remained valid after the enactment of the Federal Rules of Evidence, but the Court went beyond that issue and set forth alternative requirements for admissibility in place of the Frye test. Even though the Daubert test was announced in dicta, see 509 U.S. at 598-601 (REHNQUIST, C. J., concurring in part and dissenting in part), we should not simply ignore its analysis in reviewing the District Court's rulings.
- - - - - - - - - - - - End Footnotes- - - - - - - - - - - - - -

The relevance ruling was straightforward. The District Court correctly reasoned that an expert opinion that exposure [*152] to PCBs, "furans" and "dioxins" together may cause lung cancer would be irrelevant unless the plaintiff had been exposed to those substances. Having already found that there was no evidence of exposure to furans and dioxins, 864 F. Supp. 1310, 1318-1319 (ND Ga. 1994), it necessarily followed that this expert opinion testimony was inadmissible. Correctly applying Daubert, the District Court explained that the experts' testimony "manifestly does not fit the facts of this case, and is therefore inadmissible." 864 F. Supp. at 1322. Of course, if the evidence raised a genuine issue of fact on the question of Joiner's exposure to furans and dioxins -- as the Court of Appeals held that it did -- then this basis for the ruling on admissibility was erroneous, but not because the district judge either abused her discretion or misapplied the law. n3

- - - - - - - - - - - - - - Footnotes - - - - - - - - - - - - - - -
n3 Petitioners do not challenge the Court of Appeals' straightforward review of the District Court's summary judgment ruling on exposure to furans and dioxins. As today's opinion indicates, ante, at 10, it remains an open question on remand whether the District Court should admit expert testimony that PCBs, furans and dioxins together promoted Joiner's cancer.
- - - - - - - - - - - - End Footnotes- - - - - - - - - - - - - -

The reliability ruling was more complex and arguably is not faithful to the statement in Daubert that "the focus, of course, must be solely on principles and methodology, not on the conclusions that they generate." 509 U.S. at 595. Joiner's experts used a "weight of the evidence" [***523] methodology to assess whether Joiner's exposure to transformer fluids promoted his lung cancer. n4 They did not suggest that any [*153] one study provided adequate support for their conclusions, but instead relied on all the studies taken together (along with their interviews of Joiner and their review of his medical records). The District Court, however, examined the studies one by one and concluded that none was sufficient to show a link between PCBs and lung cancer. 864 F. Supp. at 1324-1326. The focus of the opinion was on the separate studies and the conclusions of the experts, not on the experts' methodology. Id., at 1322 ("Defendants . . . persuade the court that Plaintiffs' expert testimony would not be admissible . . . by attacking the conclusions that Plaintiffs' experts draw from the studies they cite").

- - - - - - - - - - - - - - Footnotes - - - - - - - - - - - - - - -
n4 Dr. Daniel Teitelbaum elaborated on that approach in his deposition testimony: "As a toxicologist when I look at a study, I am going to require that that study meet

the general criteria for methodology and statistical analysis, but that when all of that data is collected and you ask me as a patient, 'Doctor, have I got a risk of getting cancer from this?' That those studies don't answer the question, that I have to put them all together in my mind and look at them in relation to everything I know about the substance and everything I know about the exposure and come to a conclusion. I think when I say, 'To a reasonable medical probability as a medical toxicologist, this substance was a contributing cause,' . . . to his cancer, that that is a valid conclusion based on the totality of the evidence presented to me. And I think that that is an appropriate thing for a toxicologist to do, and it has been the basis of diagnosis for several hundred years, anyway." Supp. App. to Brief for Respondents 19.
- - - - - - - - - - - - End Footnotes- - - - - - - - - - - - - -

Unlike the District Court, the Court of Appeals expressly decided that a "weight of the evidence" methodology was scientifically acceptable. n5 To this extent, the Court of Appeals' opinion is persuasive. It is not intrinsically "unscientific" for experienced professionals to arrive at a conclusion by weighing all available scientific evidence -- this is not the sort of "junk science" with which Daubert was concerned. n6 After all, as Joiner points out, the Environmental Protection Agency (EPA) uses the same methodology to assess risks, albeit using a somewhat [**523] different threshold than that required in a trial. Brief for Respondents 40-41 (quoting [*154] EPA, Guidelines for Carcinogen Risk Assessment, 51 Fed. Reg. 33992, 33996 (1986)). Petitioners' own experts used the same scientific approach as well. n7 And using this methodology, it would seem that an expert could reasonably have concluded that the study of workers at an Italian capacitor plant, coupled with data from Monsanto's study and other studies, [***524] raises an inference that PCBs promote lung cancer. n8

- - - - - - - - - - - - - - Footnotes - - - - - - - - - - - - - - -
n5 The court explained: "Opinions of any kind are derived from individual pieces of evidence, each of which by itself might not be conclusive, but when viewed in their entirety are the building blocks of a perfectly reasonable conclusion, one reliable enough to be submitted to a jury along with the tests and criticisms cross-examination and contrary evidence would supply." 78 F.3d 524, 532 (CA11 1996).

n6 An example of "junk science" that should be excluded under Daubert as too unreliable would be the testimony of a phrenologist who would purport to prove a defendant's future dangerousness based on the contours of the defendant's skull.

n7 See, e.g., Deposition of Dr. William Charles Bailey, Supp. App. to Brief for Respondents 56 ("I've just reviewed a lot of literature and come to some conclusions . . . .").

n8 The Italian capacitor plant study found that workers exposed to PCBs had a higher-than-expected rate of lung cancer death, though "the numbers were small [and] the value of the risk estimate was not statistically significant." 864 F. Supp. 1310, 1324 (ND Ga. 1994). The Monsanto study also found a correlation between

PCB exposure and lung cancer death, but the results were not statistically significant. Id., at 1325. Moreover, it should be noted that under Georgia law, which applies in this diversity suit, Joiner need only show that his exposure to PCBs "promoted" his lung cancer, not that it was the sole cause of his cancer. Brief for Respondents 7, n. 16 (quoting Brief for Appellants in No. 94-9131 (CA 11), pp. 7-10).

- - - - - - - - - - - - End Footnotes- - - - - - - - - - - - - -

The Court of Appeals' discussion of admissibility is faithful to the dictum in Daubert that the reliability inquiry must focus on methodology, not conclusions. Thus, even though I fully agree with both the District Court's and this Court's explanation of why each of the studies on which the experts relied was by itself unpersuasive, a critical question remains unanswered: When qualified experts have reached relevant conclusions on the basis of an acceptable methodology, why are their opinions inadmissible?

Daubert quite clearly forbids trial judges from assessing the validity or strength of an expert's scientific conclusions, which is a matter for the jury. n9 Because I am persuaded [*155] that the difference between methodology and conclusions is just as categorical as the distinction between means and ends, I do not think the statement that "conclusions and methodology are not entirely distinct from one another," ante, at 9, is either accurate or helps us answer the difficult admissibility question presented by this record.

- - - - - - - - - - - - - - Footnotes - - - - - - - - - - - - - - -

n9 The Court stated in Daubert: "Vigorous cross-examination, presentation of contrary evidence, and careful instruction on the burden of proof are the traditional and appropriate means of attacking shaky but admissible evidence. . . . Additionally, in the event the trial court concludes that the scintilla of evidence presented supporting a position is insufficient to allow a reasonable juror to conclude that the position more likely than not is true, the court remains free to direct a judgment, Fed. Rule Civ. Proc. 50(a), and likewise to grant summary judgment, Fed. Rule Civ. Proc. 56. . . . These conventional devices, rather than wholesale exclusion under an uncompromising 'general acceptance' test, are the appropriate safeguards where the basis of scientific testimony meets the standards of Rule 702." 509 U.S. at 596.

- - - - - - - - - - - - End Footnotes- - - - - - - - - - - - - -

In any event, it bears emphasis that the Court has not held that it would have been an abuse of discretion to admit the expert testimony. The very point of today's holding is that the abuse of discretion standard of review applies whether the district judge has excluded or admitted evidence. Ante, at 5. And nothing in either Daubert or the Federal Rules of Evidence requires a district judge to reject an expert's conclusions and keep them from the jury when they fit the facts of the case and are based on reliable scientific methodology.

Accordingly, while I join Parts I and II of the Court's opinion, I do not concur in the judgment or in Part III of its opinion.

REFERENCES:

5 Am Jur 2d, Appellate Review 695, 700; 31A Am Jur 2d, Expert and Opinion Evidence 6

L Ed Digest, Appeal 1391, 1392; Evidence 641, 643

L Ed Index, Abuse of Discretion or Power; Experiments or Tests; Expert and Opinion Evidence; Rules of Evidence

Annotation References:

What issues will the Supreme Court consider, though not, or not properly, raised by the parties. 42 L Ed 2d 946.

Reliability of scientific technique and its acceptance within scientific community as affecting admissibility, at federal trial, of expert testimony as to result of test or study based on such technique--modern cases. 105 ALR Fed 299.

When will expert testimony "assist trier of fact" so as to be admissible at federal trial under Rule 702 of Federal Rules of Evidence. 75 ALR Fed 461.

KUMHO TIRE COMPANY, LTD., ET AL., PETITIONERS v. PATRICK CARMICHAEL, ETC., ET AL.

NO. 97-1709 [4]

SUPREME COURT OF THE UNITED STATES

526 U.S. 137; 119 S. CT. 1167; 143 L. ED. 2D 238; 1999 U.S. LEXIS 2189; 67 U.S.L.W. 4179; 50 U.S.P.Q.2D (BNA) 1177; 50 FED. R. EVID. SERV. (CALLAGHAN) 1373; CCH PROD. LIAB. REP. P15,470; 99 CAL. DAILY OP. SERVICE 2059; 29 ELR 20638; 1999 COLO. J. C.A.R. 1518; 12 FLA. L. WEEKLY FED. S 141

DECEMBER 7, 1998, ARGUED
MARCH 23, 1999, DECIDED

PRIOR HISTORY: ON WRIT OF CERTIORARI TO THE UNITED STATES COURT OF APPEALS FOR THE ELEVENTH CIRCUIT.

DISPOSITION: 131 F.3d 1433, reversed.

DECISION:

Federal trial judge's gatekeeping obligation under Federal Rules of Evidence--to insure that expert witness' testimony rests on reliable foundation and is relevant to task at hand--held to apply to all expert testimony, not only scientific.

SUMMARY:

In Daubert v Merrell Dow Pharmaceuticals, Inc. (1993) 509 US 579, 125 L Ed 2d 469, 113 S Ct 2786, a case involving the admissibility of scientific expert testimony, the United States Supreme Court held that (1) such testimony was admissible only if relevant and reliable; (2) the Federal Rules of Evidence (FRE) assigned to the trial judge the task of insuring that an expert's testimony rested on a reliable foundation and was relevant to the task at hand; and (3) some or all of certain specific factors--such as testing, peer review, error rates, and acceptability in the relevant scientific community—might possibly prove helpful in determining the reliability of a particular scientific theory or technique. In 1993, after a tire on a minivan blew out and the minivan overturned, one passenger died and the others were injured. The survivors and the decedent's representative, claiming that the failed tire had been defective, brought a diversity suit in the United States District Court for the Southern District of Alabama against the tire's maker and distributor. The plaintiffs rested their case in significant part upon the depositions of a mechanical engineer--an expert in tire failure analysis--who intended to testify that, in his expert opinion, a defect in the tire's manufacture or design caused the blowout. The expert's opinion was based upon (1) a visual and tactile inspection of the tire, and (2) the theory that in the absence of at least two of four specific physical symptoms indicating tire abuse, the tire failure of the sort that occurred in the case at hand was caused by a defect. The District Court—in granting a motion to exclude the expert's testimony as well as a motion for summary judgment against the plaintiffs--(1) agreed with the defendants that the District Court ought to act as a Daubert-type reliability "gatekeeper," even though the testimony at issue could be considered "technical" rather than "scientific"; (2) examined the expert's methodology in light of the reliability-related factors that Daubert had mentioned; and (3) concluded that all those factors argued against the reliability of the expert's methods (923 F Supp 1514, 1996 US Dist LEXIS 5706). On reconsideration, the District Court--although acknowledging that the Daubert factors ought to be applied flexibly and were simply illustrative--affirmed the earlier rulings on the ground that there were insufficient indications of the reliability of the expert's methodology of tire failure analysis. The United States Court of Appeals for the Eleventh Circuit, in reversing and remanding, expressed the view that the District Court had erred as a matter of law in applying the Daubert factors to the tire expert's testimony, as (1) Daubert was limited to the scientific context, and (2) the testimony in question relied on experience rather than the application of scientific principles (131 F3d 1433, 1997 US App LEXIS 35981).

On certiorari, the Supreme Court reversed. In an opinion by Breyer, J., joined by Rehnquist, Ch. J., and O'Connor, Scalia, Kennedy, Souter, Thomas, and Ginsburg, JJ., and joined (as to points 1 and 2 below) by Stevens, J., it was held that (1) a federal trial judge's gatekeeping obligation under the FRE—to insure that an expert witness' testimony rests on a reliable foundation and is relevant to the task at hand--applies not only to testimony based on scientific knowledge, but rather to all expert testimony, that is, testimony based on technical and other specialized knowledge; (2) in determining the admissibility of an expert's testimony--including the testimony of an engineering expert—a federal trial judge may properly consider one or more of the specific Daubert factors, where doing so will help determine that testimony's reliability; and (3) in the case at hand, the District Court's decision not to admit the expert testimony in question was within the District Court's discretion.

Scalia, J., joined by O'Connor and Thomas, JJ., concurring, expressed the view that (1) a trial court's discretion in choosing the manner of testing expert reliability is not discretion to abandon the gatekeeping function or to perform that function inadequately; and (2) in a particular case, the failure to apply one or another of the Daubert factors may possibly be unreasonable and hence an abuse of discretion.

Stevens, J., concurring in part and dissenting in part, (1) agreed that a federal trial judge may properly consider the Daubert factors in analyzing the admissibility of an engineering expert's testimony, and (2) expressed the view that the case ought to have been remanded to the Court of Appeals for a study of the record to determine whether the trial judge abused his discretion in excluding the expert testimony in question.

LAWYERS' EDITION HEADNOTES:

[***LEdHN1]
EVIDENCE §641
-- expert testimony -- judge's gatekeeping obligation
Headnote: [1A] [1B] [1C] [1D]

A federal trial judge's gatekeeping obligation under the Federal Rules of Evidence (FRE)--to insure that an expert witness' testimony rests on a reliable foundation and is relevant to the task at hand--applies not only to testimony based on scientific knowledge, but rather to all expert testimony, that is, testimony based on technical and other specialized knowledge, for (1) the language of Rule 702 of the FRE, which allows expert witnesses to give opinion testimony as to scientific, technical, or other specialized knowledge under some circumstances, (a) makes no relevant distinction between "scientific" knowledge and "technical" or "other specialized" knowledge, and (b) makes clear that any such knowledge might become the subject of expert testimony; (2) the FRE grant to all experts--not just to "scientific" ones--testimonial latitude unavailable to other witnesses on the assumption that an expert's opinion will have a reliable basis in the knowledge and experience of the expert's discipline; (3) it would prove difficult, if not impossible, for judges to administer evidentiary rules under which a gatekeeping obligation depended upon a distinction between scientific knowledge and technical or other specialized knowledge, and (4) there is no convincing need to make such distinctions.

[***LEdHN2]
EVIDENCE §641
-- expert testimony -- reliability factors
Headnote: [2A] [2B] [2C] [2D]

In determining the admissibility of an expert's testimony, including the testimony of an engineering expert, under Rule 702 of the Federal Rules of Evidence, a federal trial judge may properly consider one or more of some specific factors--whether the theory or technique (1) can be and has been tested, (2) has been subjected to peer review or publication, (3) has (a) a high known or potential rate of error, and (b) standards controlling the technique's operation, and (4) enjoys general acceptance within a relevant scientific community--where such factors are reasonable measures of the testimony's reliability; the trial judge may ask questions of this sort not only where an expert relies on the application of scientific principles, but also where an expert relies on skill- or experience-based observation.

[***LEdHN3]
EVIDENCE §643

-- expert testimony -- cause of tire failure
Headnote: [3A] [3B] [3C] [3D] [3E]

A Federal District Court's decision not to admit expert testimony as to the cause of an automobile tire's blowout is within the court's discretion, where (1) the testimony consists of the depositions of a witness who intends to testify that, in the witness' expert opinion, a defect in the tire's manufacture or design caused the tire to blow out; (2) the witness' opinion is based upon (a) a visual and tactile inspection of the tire, and (b) a theory that in the absence of at least two of four specific physical symptoms indicating tire abuse, the tire failure of the sort that occurred in the case at hand is caused by a defect; (3) the question before the court is not the reliability of the witness' methodology in general, but rather whether the witness can reliably determine the cause of failure of the particular tire at issue; (4) the witness concedes, among other matters, that this tire bore some of the very marks that were said to indicate abuse rather than a defect; (5) the witness' own testimony casts considerable doubt upon the reliability of (a) the witness' explicit theory, and (b) the implicit proposition about the significance of visual inspection in the case at hand; (6) there is no indication in the record that (a) other experts in the industry use the witness' particular approach, or (b) tire experts normally make the very fine distinctions necessary to support the witness' conclusions; (7) there are no references to articles or papers that validate the witness' approach; and (8) the court's decision is ultimately based upon the witness' failure to satisfy either (a) specific factors involving testing, peer review, error rates, and acceptability in the relevant scientific community, or (b) any other set of reasonable reliability criteria. (Stevens, J., dissented in part from this holding.)

[***LEdHN4]
EVIDENCE §641
-- expert testimony
Headnote: [4A] [4B] [4C]

For purposes of determining the admissibility, under Rule 702 of the Federal Rules of Evidence, of expert testimony that is based on a theory or technique, the test of the testimony's reliability is flexible; some specific factors that may possibly bear on the reliability determination--whether the theory or technique (1) can be and has been tested, (2) has been subjected to peer review or publication, (3) has (a) a high known or potential rate of error, and (b) standards controlling the technique's operation, and (4) enjoys general acceptance within a relevant scientific community--do not constitute a definitive checklist or test; depending on the nature of the issue, the expert's particular expertise, and the subject of the expert's testimony, such factors may or may not be pertinent in assessing the testimony's reliability; because too much depends upon the particular circumstances of the particular case at issue, the United States Supreme Court can neither rule out nor rule in the applicability of these factors (1) for all cases and for all time, or (2) for subsets of cases categorized by category of expert or by kind of evidence; these factors do not all necessarily apply in every instance in which the reliability of scientific testimony is challenged.

[***LEdHN5]
EVIDENCE §641
-- expert testimony
Headnote: [5A] [5B]

In determining the admissibility of expert testimony under Rule 702 of the Federal Rules of Evidence, a trial court must have the same kind of latitude in deciding how to test an expert's reliability--and to decide whether or when special briefing or other proceedings are needed to investigate reliability—as the trial court enjoys in deciding whether that expert's relevant testimony is reliable; thus, in determining the admissibility under Rule 702 of expert testimony that is based on a theory or technique, the question whether some specific factors--whether the theory or technique (1) can be and has been tested, (2) has been subjected to peer review or publication, (3) has (a) a high known or potential rate of error, and (b) standards controlling the technique's operation, and (4) enjoys general acceptance within a relevant scientific community--are reasonable measures of reliability in a particular case is a matter that the law grants the trial judge broad latitude to determine.

[***LEdHN6]
APPEAL §1296
-- presumptions -- expert testimony
Headnote: [6]

On certiorari to review a Federal Court of Appeals' judgment in a suit against an automobile tire's maker and distributor--in which suit an expert witness, in concluding that a defect in the tire's manufacture or design caused the tire to blow out, rests this conclusion in part upon the premises that (1) a tire's carcass should stay bound to the inner side of the tread for a significant period of time after the tread depth has worn away, (2) the tread of the tire at issue separated from the tire's inner steel-belted carcass prior to the accident, and (3) this separation caused the blowout--the United States Supreme Court must assume that these premises are not in dispute, where the witness' conclusion also rests upon some other propositions, several of which the maker and distributor dispute.

[***LEdHN7]
EVIDENCE §641
-- expert testimony
Headnote: [7]

Rule 702 of the Federal Rules of Evidence, which allows expert witnesses to give opinion testimony as to scientific, technical, or other specialized knowledge under some circumstances, establishes a standard of evidentiary reliability and requires a valid connection to the pertinent inquiry as a precondition to admissibility; where such testimony's factual basis, data, principles, methods, or their application are called sufficiently into question, the trial judge must determine whether the testimony has a reliable basis in the knowledge and experience of the relevant discipline.

[***LEdHN8]

EVIDENCE §641
-- expert testimony
Headnote: [8]

For purposes of determining the admissibility, under Rule 702 of the Federal Rules of Evidence, of expert testimony that is based on a theory or technique, the fact that the theory or technique has general acceptance within a relevant expert community does not help to show that the expert's testimony is reliable where the discipline itself lacks reliability, as, for example, theories grounded in any so-called generally accepted principles of astrology or necromancy.

[***LEdHN9]
EVIDENCE §641
-- expert testimony
Headnote: [9]

For purposes of determining the admissibility, under Rule 702 of the Federal Rules of Evidence, of expert testimony that is based on the expert's experience, (1) it is appropriate in some cases for a trial judge to ask, for example, (a) how often an engineering expert's experience-based methodology has produced erroneous results, or (b) whether such a method is generally accepted in the relevant engineering community; and (2) it is useful at times to ask even of a witness whose expertise is based purely on experience--as, for example, a perfume tester able to distinguish among 140 odors at a sniff--whether the witness' preparation is of a kind that others in the field would recognize as acceptable.

[***LEdHN10]
EVIDENCE §641
-- expert testimony
Headnote: [10]

The objective of a trial judge's gatekeeping requirement--in determining the admissibility, under Rule 702 of the Federal Rules of Evidence, of expert testimony--is to insure the reliability and relevancy of expert testimony, that is, to make certain that an expert, whether basing testimony upon professional studies or personal experience, employs in the courtroom the same level of intellectual rigor that characterizes the practice of an expert in the relevant field.

[***LEdHN11]
APPEAL §1391
-- discretion -- expert testimony
Headnote: [11A] [11B]

A Federal Court of Appeals is to apply an abuse-of-discretion standard when reviewing a federal trial court's decision to admit or exclude expert testimony; this standard applies as much to the trial court's decisions about how to determine reliability as to the trial court's ultimate conclusion, for otherwise, the trial judge would lack the discretionary authority needed to (1) avoid unnecessary reliability proceedings in ordinary cases where the reliability of an expert's methods is properly taken for granted, and (2) require appropriate proceedings in the less usual or more complex cases where cause for questioning the expert's reliability arises.

[***LEdHN12]
EVIDENCE §641
-- expert testimony
Headnote: [12]

Nothing in a United States Supreme Court decision involving the admissibility of expert testimony or in the Federal Rules of Evidence requires a Federal District Court to admit opinion evidence that is connected to existing data by only the expert's own statement.

SYLLABUS: When a tire on the vehicle driven by Patrick Carmichael blew out and the vehicle overturned, one passenger died and the others were injured. The survivors and the decedent's representative, respondents here, brought this diversity suit against the tire's maker and its distributor (collectively Kumho Tire), claiming that the tire that failed was defective. They rested their case in significant part upon the depositions of a tire failure analyst, Dennis Carlson, Jr., who intended to testify that, in his expert opinion, a defect in the tire's manufacture or design caused the blow out. That opinion was based upon a visual and tactile inspection of the tire and upon the theory that in the absence of at least two of four specific, physical symptoms indicating tire abuse, the tire failure of the sort that occurred here was caused by a defect. Kumho Tire moved to exclude Carlson's testimony on the ground that his methodology failed to satisfy Federal Rule of Evidence 702, which says: "If scientific, technical, or other specialized knowledge will assist the trier of fact . . . , a witness qualified as an expert . . . may testify thereto in the form of an opinion." Granting the motion (and entering summary judgment for the defendants), the District Court acknowledged that it should act as a reliability "gatekeeper" under Daubert v. Merrell Dow Pharmaceuticals, Inc., 509 U.S. 579, 589, 125 L. Ed. 2d 469, 113 S. Ct. 2786, in which this Court held that Rule 702 imposes a special obligation upon a trial judge to ensure that scientific testimony is not only relevant, but reliable. The court noted that Daubert discussed four factors -- testing, peer review, error rates, and "acceptability" in the relevant scientific community -- which might prove helpful in determining the reliability of a particular scientific theory or technique, 509 U.S. at 593-594, and found that those factors argued against the reliability of Carlson's methodology. On the plaintiffs' motion for reconsideration, the court agreed that Daubert should be applied flexibly, that its four factors were simply illustrative, and that other factors could argue in favor of admissibility. However, the court affirmed its earlier order because it found insufficient indications of the reliability of Carlson's methodology. In reversing, the Eleventh Circuit held that the District Court had erred as a matter of law in applying Daubert. Believing that Daubert was limited to the scientific context, the court held that the Daubert factors did not apply to Carlson's testimony, which it characterized as skill- or experience-based.

Held:

1. The Daubert factors may apply to the testimony of engineers and other experts who are not scientists. Pp. 7-13.

(a) The Daubert "gatekeeping" obligation applies not only to "scientific" testimony, but to all expert testimony. Rule 702 does not distinguish between "scientific" knowledge and "technical" or "other specialized" knowledge, but makes clear that any such knowledge might become the subject of expert testimony. It is the Rule's word "knowledge," not the words (like "scientific") that modify that word, that establishes a standard of evidentiary reliability. 509 U.S. at 589-590. Daubert referred only to "scientific" knowledge because that was the nature of the expertise there at issue. Id. at 590, n. 8. Neither is the evidentiary rationale underlying Daubert's "gatekeeping" determination limited to "scientific" knowledge. Rules 702 and 703 grant all expert witnesses, not just "scientific" ones, testimonial latitude unavailable to other witnesses on the assumption that the expert's opinion will have a reliable basis in the knowledge and experience of his discipline. Id. at 592. Finally, it would prove difficult, if not impossible, for judges to administer evidentiary rules under which a "gatekeeping" obligation depended upon a distinction between "scientific" knowledge and "technical" or "other specialized" knowledge, since there is no clear line dividing the one from the others and no convincing need to make such distinctions. Pp. 7-9.

(b) A trial judge determining the admissibility of an engineering expert's testimony may consider one or more of the specific Daubert factors. The emphasis on the word "may" reflects Daubert's description of the Rule 702 inquiry as "a flexible one." 509 U.S. at 594. The Daubert factors do not constitute a definitive checklist or test, id. at 593, and the gatekeeping inquiry must be tied to the particular facts, id. at 591. Those factors may or may not be pertinent in assessing reliability, depending on the nature of the issue, the expert's particular expertise, and the subject of his testimony. Some of those factors may be helpful in evaluating the reliability even of experience-based expert testimony, and the Court of Appeals erred insofar as it ruled those factors out in such cases. In determining whether particular expert testimony is reliable, the trial court should consider the specific Daubert factors where they are reasonable measures of reliability. Pp. 10-12.

(c) The court of appeals must apply an abuse-of-discretion standard when it reviews the trial court's decision to admit or exclude expert testimony. General Electric Co. v. Joiner, 522 U.S. 136, 138-139, 139 L. Ed. 2d 508, 118 S. Ct. 512. That standard applies as much to the trial court's decisions about how to determine reliability as to its ultimate conclusion. Thus, whether Daubert's specific factors are, or are not, reasonable measures of reliability in a particular case is a matter that the law grants the trial judge broad latitude to determine. See id. at 143. The Eleventh Circuit erred insofar as it held to the contrary. P. 13.

2. Application of the foregoing standards demonstrates that the District Court's decision not to admit Carlson's expert testimony was lawful. The District Court did not question Carlson's qualifications, but excluded his testimony because it initially

doubted his methodology and then found it unreliable after examining the transcript in some detail and considering respondents' defense of it. The doubts that triggered the court's initial inquiry were reasonable, as was the court's ultimate conclusion that Carlson could not reliably determine the cause of the failure of the tire in question. The question was not the reliability of Carlson's methodology in general, but rather whether he could reliably determine the cause of failure of the particular tire at issue. That tire, Carlson conceded, had traveled far enough so that some of the tread had been worn bald, it should have been taken out of service, it had been repaired (inadequately) for punctures, and it bore some of the very marks that he said indicated, not a defect, but abuse. Moreover, Carlson's own testimony cast considerable doubt upon the reliability of both his theory about the need for at least two signs of abuse and his proposition about the significance of visual inspection in this case. Respondents stress that other tire failure experts, like Carlson, rely on visual and tactile examinations of tires. But there is no indication in the record that other experts in the industry use Carlson's particular approach or that tire experts normally make the very fine distinctions necessary to support his conclusions, nor are there references to articles or papers that validate his approach. Respondents' argument that the District Court too rigidly applied Daubert might have had some validity with respect to the court's initial opinion, but fails because the court, on reconsideration, recognized that the relevant reliability inquiry should be "flexible," and ultimately based its decision upon Carlson's failure to satisfy either Daubert's factors or any other set of reasonable reliability criteria. Pp. 13-19.

131 F.3d 1433, reversed.

COUNSEL:

Joseph H. Babington argued the cause for petitioners.

Jeffrey P. Minear argued the cause for the United States, as amicus curiae, by special leave of court.

Sidney W. Jackson argued the cause for respondents.

JUDGES: BREYER, J., delivered the opinion of the Court, in which REHNQUIST, C.J., and O'CONNOR, SCALIA, KENNEDY, SOUTER, THOMAS, and GINSBURG, JJ., joined, and in which STEVENS, J., joined as to Parts I and II. SCALIA, J., filed a concurring opinion, in which O'CONNOR and THOMAS, JJ., joined. STEVENS, J., filed an opinion concurring in part and dissenting in part.

OPINIONBY: BREYER

OPINION: [*141] [**1171] [***246] JUSTICE BREYER delivered the opinion of the Court.

[***LEdHR1A] [1A] In Daubert v. Merrell Dow Pharmaceuticals, Inc., 509 U.S. 579, 125 L. Ed. 2d 469, 113 S. Ct. 2786 (1993), this Court focused upon the admissibility of scientific expert testimony. It pointed out that such testimony is admissible only if it is both relevant and reliable. And it held that the Federal Rules of Evidence "assign to the trial judge the task of ensuring that an expert's testimony both rests on a reliable foundation and is relevant to the task at hand." Id. at 597. The Court also discussed certain more specific factors, such as testing, peer review, error rates, and "acceptability" in the relevant scientific community, some or all of which might prove helpful in determining the reliability of a particular scientific "theory or technique." 509 U.S. at 593-594.

[***LEdHR1B] [1B] [***LEdHR2A] [2A] [***LEdHR3A] [3A] [***LEdHR4A] [4A]
[***LEdHR5A] [5A] This case requires us to decide how Daubert applies to the testimony of engineers and other experts who are not scientists. We conclude that Daubert's general holding -- setting forth the trial judge's general "gatekeeping" obligation -- applies not only to testimony based on "scientific" knowledge, but also to testimony based on "technical" and "other specialized" knowledge. See Fed. Rule Evid. 702. We also conclude that a trial court may consider one or more of the more specific factors that Daubert mentioned when doing so will help determine that testimony's reliability. But, as the Court stated in Daubert, the test of reliability is "flexible," and Daubert's list of specific factors neither necessarily nor exclusively applies to all experts or in every case. [*142] Rather, the law grants a district court the same broad latitude when it decides how to determine reliability as it enjoys in respect to its ultimate reliability determination. See General Electric Co. v. Joiner, 522 U.S. 136, 143, 139 L. Ed. 2d 508, 118 S. Ct. 512 (1997) (courts of [***247] appeals are to apply "abuse of discretion" standard when reviewing district court's reliability determination). Applying these standards, we determine that the District Court's decision in this case -- not to admit certain expert testimony -- was within its discretion and therefore lawful.

I

On July 6, 1993, the right rear tire of a minivan driven by Patrick Carmichael blew out. In the accident that followed, one of the passengers died, and others were severely injured. In October 1993, the Carmichaels brought this diversity suit against the tire's maker and its distributor, whom we refer to collectively as Kumho Tire, claiming that the tire was defective. The plaintiffs rested their case in significant part upon deposition testimony provided by an expert in tire failure analysis, Dennis Carlson, Jr., who intended to testify in support of their conclusion.

Carlson's depositions relied upon certain features of tire technology that are not in dispute. A steel-belted radial tire like the Carmichaels' is made up of a "carcass" containing many layers of flexible cords, called "plies," along which (between the cords and the outer tread) are laid steel strips called "belts." Steel wire loops, called "beads," hold the cords together at the plies' bottom edges. An outer layer, called the "tread," encases the carcass, and the entire tire is bound together in rubber, through the application of heat and various chemicals. See generally, e.g., J. Dixon, Tires, Suspension and Handling 68-72 (2d ed. 1996). The bead of the tire sits upon a "bead seat," which is part of the wheel assembly. That assembly contains a "rim flange," which extends over the bead and rests against the side of the [*143] tire. See M. Mavrigian, Performance Wheels & Tires 81, 83 (1998) (illustrations).

[Graphic omitted; see printed opinion.]

A. Markovich, How To Buy and Care For Tires 4 (1994).

[**1172] Carlson's testimony also accepted certain background facts about the tire in question. He assumed that before the blowout the tire had traveled far. (The tire was made in 1988 and had been installed some time before the Carmichaels bought the used minivan in March 1993; the Carmichaels had driven the van approximately 7,000 additional miles in the two months they had owned it.) Carlson noted that the tire's tread depth, which was 11/32 of an inch when new, App. 242, had been worn down to depths that ranged from 3/32 of an inch along some parts of the tire, to nothing at all along others. Id. at 287. He conceded that the tire tread had at least two punctures which had been inadequately repaired. Id. at 258-261, 322.

[***LEdHR6] [6]Despite the tire's age and history, Carlson concluded that a defect in its manufacture or design caused the blow-out. He rested this conclusion in part upon three premises which, [*144] for present purposes, we must assume are not in dispute: First, a tire's carcass should stay [***248] bound to the inner side of the tread for a significant period of time after its tread depth has worn away. Id. at 208-209. Second, the tread of the tire at issue had separated from its inner steel-belted carcass prior to the accident. Id. at 336. Third, this "separation" caused the blowout. Ibid.

Carlson's conclusion that a defect caused the separation, however, rested upon certain other propositions, several of which the defendants strongly dispute. First, Carlson said that if a separation is not caused by a certain kind of tire misuse called "overdeflection" (which consists of underinflating the tire or causing it to carry too much weight, thereby generating heat that can undo the chemical tread/carcass bond), then, ordinarily, its cause is a tire defect. Id. at 193-195, 277-278. Second, he said that if a tire has been subject to sufficient overdeflection to cause a separation, it should reveal certain physical symptoms. These symptoms include (a) tread wear on the tire's shoulder that is greater than the tread wear along the tire's center, id. at 211; (b) signs of a "bead groove," where the beads have been pushed too hard against the bead seat on the inside of the tire's rim, id. at 196-197; (c) sidewalls of the tire with physical signs of deterioration, such as discoloration, id. at 212; and/or (d) marks on the tire's rim flange, id. at 219-220. Third, Carlson said that where he does not find at least two of the four physical signs just mentioned (and presumably where there is no reason to suspect a less common cause of separation), he concludes that a manufacturing or design defect caused the separation. Id. at 223-224.

Carlson added that he had inspected the tire in question. He conceded that the tire to a limited degree showed greater wear on [**1173] the shoulder than in the center, some signs of "bead groove," some discoloration, a few marks on the rim flange, and inadequately filled puncture holes (which can also cause heat that might lead to separation). Id. at 256-257, 258-261, [*145] 277, 303-304, 308. But, in each instance, he testified that the symptoms were not significant, and he explained why he believed that they did not reveal overdeflection. For example, the extra shoulder wear, he said, appeared primarily on one shoulder, whereas an overdeflected tire would reveal equally abnormal wear on both shoulders. Id. at 277. Carlson concluded that the tire did not bear at least two of the four overdeflection symptoms, nor was there any less obvious cause of separation; and since neither overdeflection nor the punctures caused the blowout, a defect must have done so.

Kumho Tire moved the District Court to exclude Carlson's testimony on the ground that his methodology failed Rule 702's reliability requirement. The court agreed with Kumho that it should act as a Daubert-type reliability "gatekeeper," even though one might consider Carlson's testimony as "technical," rather than "scientific." See Carmichael v. Samyang Tires, Inc., 923 F. Supp. 1514, 1521-1522 (SD Ala. 1996). The court then examined Carlson's methodology in light of the reliability-related factors that Daubert mentioned, such as a theory's testability, whether it "has been a subject of peer review or publication," the "known or potential rate of error," and the "degree of acceptance . . . within the relevant scientific community." 923 F. Supp. at 1520 (citing Daubert, 509 U.S. 579 at 592-594). [***249] The District Court found that all those factors argued against the reliability of Carlson's methods, and it granted the motion to exclude the testimony (as well as the defendants' accompanying motion for summary judgment).

The plaintiffs, arguing that the court's application of the Daubert factors was too "inflexible," asked for reconsideration. And the Court granted that motion. Carmichael v. Samyang Tires, Inc., 1996 U.S. Dist. LEXIS 22431, Civ. Action No. 93-0860-CB-S (June 5, 1996), App. to Pet. for Cert. 1c. After reconsidering the matter, the court agreed with the plaintiffs that Daubert should be applied flexibly, that its four factors were [*146] simply illustrative, and that other factors could argue in favor of admissibility. It conceded that there may be widespread acceptance of a "visual-inspection method" for some relevant purposes. But the court found insufficient indications of the reliability of "the component of Carlson's tire failure analysis which most concerned the Court, namely, the methodology employed by the expert in analyzing the data obtained in the visual inspection, and the scientific basis, if any, for such an analysis." Id. at 6c.

It consequently affirmed its earlier order declaring Carlson's testimony inadmissible and granting the defendants' motion for summary judgment.

The Eleventh Circuit reversed. See Carmichael v. Samyang Tire, Inc., 131 F.3d 1433 (1997). It "reviewed . . . de novo" the "district court's legal decision to apply Daubert." 131 F.3d at 1435. It noted that "the Supreme Court in Daubert explicitly limited its holding to cover only the 'scientific context,'" adding that "a Daubert analysis" applies only where an expert relies "on the application of scientific principles," rather than "on skill- or experience-based observation." 131 F.3d at 1435-1436. It concluded that Carlson 's testimony, which it viewed as relying on experience, "falls outside the scope of Daubert," that "the district court erred as a matter of law by applying Daubert in this case," and that the case must be remanded for further (non-Daubert-type) consideration under Rule 702. Id. at 1436.

Kumho Tire petitioned for certiorari, asking us to determine whether a trial court "may" consider Daubert's specific "factors" when determining the "admissibility of an engineering expert's testimony." Pet. for Cert. i. We granted certiorari in light of uncertainty among the lower courts about whether, or how, Daubert applies to expert testimony that might be characterized as based not upon "scientific" knowledge, but rather upon "technical" or "other specialized" [*147] knowledge. Fed. Rule Evid. 702; compare, e.g., Watkins v. Telsmith, Inc., 121 F.3d 984, 990-991 (CA5 1997), with, e.g., Compton v. Subaru of America, Inc., 82 F.3d 1513, 1518-1519 [**1174] (CA10), cert. denied, 519 U.S. 1042, 136 L. Ed. 2d 536, 117 S. Ct. 611 (1996).

II

A

[***LEdHR1C] [1C] In Daubert, this Court held that Federal Rule of Evidence 702 imposes a special obligation upon a trial judge to "ensure that any and all scientific testimony . . . is not only relevant, but reliable." 509 U.S. at 589. The initial question before us is whether this basic gatekeeping obligation applies only to "scientific" [***250] testimony or to all expert testimony. We, like the parties, believe that it applies to all expert testimony. See Brief for Petitioners 19; Brief for Respondents 17.

For one thing, Rule 702 itself says:

"If scientific, technical, or other specialized knowledge will assist the trier of fact to understand the evidence or to determine a fact in issue, a witness qualified as an expert by knowledge, skill, experience, training, or education, may testify thereto in the form of an opinion or otherwise."

This language makes no relevant distinction between "scientific" knowledge and "technical" or "other specialized" knowledge. It makes clear that any such knowledge might become the subject of expert testimony. In Daubert, the Court specified that it is the Rule's word "knowledge," not the words (like "scientific") that modify that word, that "establishes a standard of evidentiary reliability." 509 U.S. at 589-590. Hence, as a matter of language, the Rule applies its reliability standard to all "scientific," "technical," or "other specialized" matters within its scope. We concede that the Court in Daubert referred only to "scientific" knowledge. But as the Court there said, it referred to "scientific" [*148] testimony "because that was the nature of the expertise" at issue. 509 U.S. at 590, n.8.

Neither is the evidentiary rationale that underlay the Court's basic Daubert "gatekeeping" determination limited to "scientific" knowledge. Daubert pointed out that Federal Rules 702 and 703 grant expert witnesses testimonial latitude unavailable to other witnesses on the "assumption that the expert's opinion will have a reliable basis in the knowledge and experience of his discipline." 509 U.S. at 592 (pointing out that experts may testify to opinions, including those that are not based on firsthand knowledge or observation). The Rules grant that latitude to all experts, not just to "scientific" ones.

Finally, it would prove difficult, if not impossible, for judges to administer evidentiary rules under which a gatekeeping obligation depended upon a distinction between "scientific" knowledge and "technical" or "other specialized" knowledge. There is no clear line that divides the one from the others. Disciplines such as engineering rest upon scientific knowledge. Pure scientific theory itself may depend for its development upon observation and properly engineered machinery. And conceptual efforts to distinguish the two are unlikely to produce clear legal lines capable of application in particular cases. Cf. Brief for National Academy of

Engineering as Amicus Curiae 9 (scientist seeks to understand nature while the engineer seeks nature's modification); Brief for Rubber Manufacturers Association as Amicus Curiae 14-16 (engineering, as an "applied science," relies on "scientific reasoning and methodology"); Brief for John Allen et al. as Amici Curiae 6 (engineering relies upon "scientific knowledge and methods").

Neither is there a convincing need to make such distinctions. Experts of all kinds tie observations to conclusions through the use of what Judge Learned Hand called "general truths derived from . . . specialized experience." Hand, Historical and Practical Considerations Regarding Expert Testimony, [*149] 15 Harv. L. Rev. 40, 54 (1901). And whether the specific [***251] expert testimony focuses upon specialized observations, the specialized translation of those observations into theory, a specialized theory itself, or the application of such a theory in a particular case, the expert's testimony often will rest "upon an experience confessedly foreign in kind to [the jury's] own." Ibid. The trial judge's effort to assure that the specialized testimony is reliable and relevant can help the jury evaluate [**1175] that foreign experience, whether the testimony reflects scientific, technical, or other specialized knowledge.

[***LEdHR1D] [1D] [***LEdHR7] [7]We conclude that Daubert's general principles apply to the expert matters described in Rule 702. The Rule, in respect to all such matters, "establishes a standard of evidentiary reliability." 509 U.S. at 590. It "requires a valid . . . connection to the pertinent inquiry as a precondition to admissibility." 509 U.S. at 592. And where such testimony's factual basis, data, principles, methods, or their application are called sufficiently into question, see Part III, infra, the trial judge must determine whether the testimony has "a reliable basis in the knowledge and experience of [the relevant] discipline." 509 U.S. at 592.

B [***LEdHR2B] [2B]The petitioners ask more specifically whether a trial judge determining the "admissibility of an engineering expert's testimony" may consider several more specific factors that Daubert said might "bear on" a judge's gate-keeping determination. These factors include:

-- Whether a "theory or technique . . . can be (and has been) tested";

-- Whether it "has been subjected to peer review and publication";

-- Whether, in respect to a particular technique, there is a high "known or potential rate of error" and whether there are "standards controlling the technique's operation"; and [*150]

-- Whether the theory or technique enjoys "general acceptance" within a "relevant scientific community." 509 U.S. at 592-594.

Emphasizing the word "may" in the question, we answer that question yes.

[***LEdHR4B] [4B]Engineering testimony rests upon scientific foundations, the reliability of which will be at issue in some cases. See, e.g., Brief for Stephen Bobo et al. as Amici Curiae 23 (stressing the scientific bases of engineering disciplines). In other cases, the relevant reliability concerns may focus upon personal knowledge or experience. As the Solicitor General points out, there are many different kinds of experts, and many different kinds of expertise. See Brief for United States as

Amicus Curiae 18-19, and n. 5 (citing cases involving experts in drug terms, handwriting analysis, criminal modus operandi, land valuation, agricultural practices, railroad procedures, attorney's fee valuation, and others). Our emphasis on the word "may" thus reflects Daubert's description of the Rule 702 inquiry as "a flexible one." 509 U.S. at 594. Daubert makes clear that the factors it mentions do not constitute a "definitive checklist or test." 509 U.S. at 593. And Daubert adds that the gatekeeping inquiry must be "'tied to the facts'" of a particular "case." 509 U.S. at 591 (quoting United States v. Downing, 753 F.2d 1224, 1242 (CA3 1985)). We agree with the Solicitor General that "the factors identified in Daubert may or may not be pertinent in assessing reliability, depending [***252] on the nature of the issue, the expert's particular expertise, and the subject of his testimony." Brief for United States as Amicus Curiae 19. The conclusion, in our view, is that we can neither rule out, nor rule in, for all cases and for all time the applicability of the factors mentioned in Daubert, nor can we now do so for subsets of cases categorized by category of expert or by kind of evidence. Too much depends upon the particular circumstances of the particular case at issue. [*151] [***LEdHR4C] [4C] [***LEdHR8] [8]Daubert itself is not to the contrary. It made clear that its list of factors was meant to be helpful, not definitive. Indeed, those factors do not all necessarily apply even in every instance in which the reliability of scientific testimony is challenged. It might not be surprising in a particular case, for example, that a claim made by a scientific witness has never been the subject of peer review, for the particular application at issue may never previously have interested any scientist. Nor, on the other hand, does the presence of Daubert's general acceptance factor help show that an expert's testimony is reliable where the discipline itself lacks reliability, as, for example, do theories grounded in any so-called generally accepted principles of astrology or necromancy.

[**1176]

[***LEdHR9] [9]At the same time, and contrary to the Court of Appeals' view, some of Daubert's questions can help to evaluate the reliability even of experience-based testimony. In certain cases, it will be appropriate for the trial judge to ask, for example, how often an engineering expert's experience-based methodology has produced erroneous results, or whether such a method is generally accepted in the relevant engineering community. Likewise, it will at times be useful to ask even of a witness whose expertise is based purely on experience, say, a perfume tester able to distinguish among 140 odors at a sniff, whether his preparation is of a kind that others in the field would recognize as acceptable.

[***LEdHR2C] [2C]We must therefore disagree with the Eleventh Circuit's holding that a trial judge may ask questions of the sort Daubert mentioned only where an expert "relies on the application of scientific principles," but not where an expert relies "on skill- or experience-based observation." 131 F.3d at 1435. We do not believe that Rule 702 creates a schematism that segregates expertise by type while mapping certain kinds of questions to certain kinds of experts. Life and the legal cases that it generates are too complex to warrant so definitive a match.

[*152] [***LEdHR2D] [2D] [***LEdHR10] [10]To say this is not to deny the importance of Daubert's gatekeeping requirement. The objective of that requirement is to ensure the reliability and relevancy of expert testimony. It is to make certain that an expert, whether basing testimony upon professional studies or personal experience, employs in the courtroom the same level of intellectual rigor that characterizes the practice of an expert in the relevant field. Nor do we deny that, as stated in Daubert, the particular questions that it mentioned will often be appropriate for use in determining the reliability of challenged expert testimony. Rather, we conclude that the trial judge must have considerable leeway in deciding in a particular case how to go about determining whether particular expert testimony is reliable. That is to say, a trial court should consider the specific factors identified in Daubert where they are reasonable measures of the reliability of expert testimony.

C

[***LEdHR5B] [5B] [***LEdHR11A] [11A] The trial court must have the same kind of latitude in deciding how to test an expert's reliability, and to decide whether or when special briefing or other proceedings are [***253] needed to investigate reliability, as it enjoys when it decides whether that expert's relevant testimony is reliable. Our opinion in Joiner makes clear that a court of appeals is to apply an abuse-of-discretion standard when it "reviews a trial court's decision to admit or exclude expert testimony." 522 U.S. at 138-139. That standard applies as much to the trial court's decisions about how to determine reliability as to its ultimate conclusion. Otherwise, the trial judge would lack the discretionary authority needed both to avoid unnecessary "reliability" proceedings in ordinary cases where the reliability of an expert's methods is properly taken for granted, and to require appropriate proceedings in the less usual or more complex cases where cause for questioning the expert's reliability arises. Indeed, the Rules seek to avoid "unjustifiable expense and delay" as part of their search for [*153] "truth" and the "just determination" of proceedings. Fed. Rule Evid. 102. Thus, whether Daubert's specific factors are, or are not, reasonable measures of reliability in a particular case is a matter that the law grants the trial judge broad latitude to determine. See Joiner, supra, at 143. And the Eleventh Circuit erred insofar as it held to the contrary.

III

[***LEdHR3B] [3B] We further explain the way in which a trial judge "may" consider Daubert's factors by applying these considerations to the case at hand, a matter that has been briefed exhaustively by the parties and their 19 amici. The District Court did not doubt Carlson's qualifications, which included a masters degree in mechanical engineering, 10 years' work at Michelin America, Inc., and testimony as a tire failure consultant in other tort cases. Rather, it excluded the testimony because, despite those qualifications, it initially [**1177] doubted, and then found unreliable, "the methodology employed by the expert in analyzing the data obtained in the visual inspection, and the scientific basis, if any, for such an analysis." Civ. Action No. 93-0860-CB-S (SD Ala., June 5, 1996), App. to Pet. for Cert. 6c. After examining the transcript in "some detail," 923 F. Supp. at 1518-519, n. 4, and after considering respondents' defense of Carlson's methodology, the District Court determined that Carlson's testimony was not reliable. It fell outside the range where experts might reasonably differ, and where the jury must decide among the conflicting views of different experts, even though the evidence is "shaky." Daubert, 509 U.S. at 596. In our view, the doubts that triggered the District Court's initial inquiry here were reasonable, as was the court's ultimate conclusion.

For one thing, and contrary to respondents' suggestion, the specific issue before the court was not the reasonableness in general of a tire expert's use of a visual and tactile inspection to determine whether overdeflection had caused [*154] the tire's tread to separate from its steel-belted carcass. Rather, it was the reasonableness of using such an approach, along with Carlson's particular method of analyzing the data thereby obtained, to draw a conclusion regarding the particular matter to which the expert testimony was directly relevant. That matter concerned the likelihood that a defect in the tire at issue caused its tread to separate from its carcass. The tire in question, the expert conceded, had traveled far enough so that some of the tread had been worn bald; it should have been taken out of service; it had been repaired (inadequately) for punctures; and it bore some of the very marks that the [***254] expert said indicated, not a defect, but abuse through overdeflection. See supra, at 3-5; App. 293-294. The relevant issue was whether the expert could reliably determine the cause of this tire's separation. Nor was the basis for Carlson's conclusion simply the general theory that, in the absence of evidence of abuse, a defect will normally have caused a tire's separation. Rather, the expert employed a more specific theory to establish the existence (or absence) of such abuse. Carlson testified precisely that in the absence of at least two of four signs of abuse (proportionately greater tread wear on the shoulder; signs of grooves caused by the beads; discolored sidewalls; marks on the rim flange) he concludes that a defect caused the separation. And his analysis depended upon acceptance of a further implicit proposition, namely, that his visual and tactile inspection could determine that the tire before him had not been abused despite some evidence of the presence of the very signs for which he looked (and two punctures).

For another thing, the transcripts of Carlson's depositions support both the trial court's initial uncertainty and its final conclusion. Those transcripts cast considerable doubt upon the reliability of both the explicit theory (about the need for two signs of abuse) and the implicit proposition (about the significance of visual inspection in this case). Among other things, the expert could not say whether the tire had traveled [*155] more than 10, or 20, or 30, or 40, or 50 thousand miles, adding that 6,000 miles was "about how far" he could "say with any certainty." Id. at 265. The court could reasonably have wondered about the reliability of a method of visual and tactile inspection sufficiently precise to ascertain with some certainty the abuse-related significance of minute shoulder/center relative tread wear differences, but insufficiently precise to tell "with any certainty" from the tread wear whether a tire had traveled less than 10,000 or more than 50,000 miles. And these concerns might have been augmented by Carlson's repeated reliance on the "subjectiveness" of his mode of analysis in response to questions seeking specific information regarding how he could differentiate between a tire that actually had been overdeflected and a tire that merely looked as though it had been. Id. at 222, 224-225, 285-286. They would have been further augmented by the fact that Carlson said he had inspected the tire itself for the first time the morning of his first deposition, and then only for a few hours. (His initial conclusions were based on photographs.) Id. at 180.

[**1178] Moreover, prior to his first deposition, Carlson had issued a signed report in which he concluded that the tire had "not been . . . overloaded or underinflated," not because of the absence of "two of four" signs of abuse, but simply because "the rim flange impressions . . . were normal." Id. at 335-336. That report also said that the "tread depth remaining was 3/32 inch," id. at 336, though the opposing expert's (apparently undisputed) measurements indicate that the tread depth taken at various positions around the tire actually ranged from .5/32 of an inch to 4/32 of an inch, with the tire apparently showing greater wear along both shoulders than along the center, id. at 432-433.

Further, in respect to one sign of [***255] abuse, bead grooving, the expert seemed to deny the sufficiency of his own simple visual-inspection methodology. He testified that most tires have some bead groove pattern, that where there is reason [*156] to suspect an abnormal bead groove he would ideally "look at a lot of [similar] tires" to know the grooving's significance, and that he had not looked at many tires similar to the one at issue. Id. at 212-213, 214, 217.

Finally, the court, after looking for a defense of Carlson's methodology as applied in these circumstances, found no convincing defense. Rather, it found (1) that "none" of the Daubert factors, including that of "general acceptance" in the relevant expert community, indicated that Carlson's testimony was reliable, 923 F. Supp. at 1521; (2) that its own analysis "revealed no countervailing factors operating in favor of admissibility which could outweigh those identified in Daubert," App. to Pet. for Cert. 4c; and (3) that the "parties identified no such factors in their briefs," ibid. For these three reasons taken together, it concluded that Carlson's testimony was unreliable.

Respondents now argue to us, as they did to the District Court, that a method of tire failure analysis that employs a visual/tactile inspection is a reliable method, and they point both to its use by other experts and to Carlson's long experience working for Michelin as sufficient indication that that is so. But no one denies that an expert might draw a conclusion from a set of observations based on extensive and specialized experience. Nor does anyone deny that, as a general matter, tire abuse may often be identified by qualified experts through visual or tactile inspection of the tire. See Affidavit of H. R. Baumgardner 1-2, cited in Brief for National Academy of Forensic Engineers as Amici Curiae 16 (Tire engineers rely on visual examination and process of elimination to analyze experimental test tires). As we said before, supra, at 14, the question before the trial court was specific, not general. The trial court had to decide whether this particular expert had sufficient specialized knowledge to assist the jurors "in deciding the particular issues in the case." 4 J. McLaughlin, Weinstein's Federal Evidence P702.05[1], p. 702-33 (2d ed. 1998); see also Advisory [*157] Committee's Note on Proposed Fed. Rule Evid. 702, Preliminary Draft of Proposed Amendments to the Federal Rules of Civil Procedure and Evidence: Request for Comment 126 (1998) (stressing that district courts must "scrutinize" whether the "principles and methods" employed by an expert "have been properly applied to the facts of the case").

[***LEdHR3C] [3C] [***LEdHR12] [12]The particular issue in this case concerned the use of Carlson's two-factor test and his related use of visual/tactile inspection to draw conclusions on the basis of what seemed small observational differences. We have found no indication in the record that other experts in the industry use Carlson's two-factor test or that tire experts such as Carlson normally make the very fine distinctions about, say, the symmetry of comparatively greater shoulder tread wear that were necessary, on Carlson's own theory, to support his conclusions. Nor, despite the prevalence of tire testing, does anyone refer to any articles or papers that validate Carlson's approach. Compare Bobo, Tire Flaws and Separations, in Mechanics of Pneumatic Tires 636-637 (S. Clark ed. 1981); C. Schnuth et al., Compression Grooving and Rim Flange Abrasion [***256] as Indicators of Over-Deflected Operating Conditions in Tires, presented to Rubber Division of the American Chemical Society, Oct. 21-24, 1997; J. Walter & R. Kiminecz, Bead [**1179] Contact Pressure Measurements at the Tire-Rim Interface, presented to Society of Automotive Engineers, Feb. 24-28, 1975. Indeed, no one has argued that Carlson himself, were he still working for Michelin, would have concluded in a report to his employer that a similar tire was similarly defective on grounds identical to those upon which he rested his conclusion here. Of course, Carlson himself claimed that his method was accurate, but, as we pointed out in Joiner, nothing in either Daubert or the Federal Rules of Evidence requires a district court to admit opinion evidence that is connected to existing data only by the ipse dixit of the expert." 522 U.S. at 146.

[*158] [***LEdHR3D] [3D]Respondents additionally argue that the District Court too rigidly applied Daubert's criteria. They read its opinion to hold that a failure to satisfy any one of those criteria automatically renders expert testimony inadmissible. The District Court's initial opinion might have been vulnerable to a form of this argument. There, the court, after rejecting respondents' claim that Carlson's testimony was "exempted from Daubert-style scrutiny" because it was "technical analysis" rather than "scientific evidence," simply added that "none of the four admissibility criteria outlined by the Daubert court are satisfied." 923 F. Supp. at 1522. Subsequently, however, the court granted respondents' motion for reconsideration. It then explicitly recognized that the relevant reliability inquiry "should be 'flexible,'" that its "'overarching subject [should be] . . . validity' and reliability," and that "Daubert was intended neither to be exhaustive nor to apply in every case." App. to Pet. for Cert. 4c (quoting Daubert, 509 U.S. at 594-595). And the court ultimately based its decision upon Carlson's failure to satisfy either Daubert's factors or any other set of reasonable reliability criteria. In light of the record as developed by the parties, that conclusion was within the District Court's lawful discretion.

[***LEdHR3E] [3E] [***LEdHR11B] [11B]In sum, Rule 702 grants the district judge the discretionary authority, reviewable for its abuse, to determine reliability in light of the particular facts and circumstances of the particular case. The District Court did not abuse its discretionary authority in this case. Hence, the judgment of the Court of Appeals is Reversed.

CONCURBY: SCALIA

CONCUR:

JUSTICE SCALIA, with whom JUSTICE O'CONNOR and JUSTICE THOMAS join, concurring.

I join the opinion of the Court, which makes clear that the discretion it endorses-trial-court discretion in choosing the manner of testing expert reliability- is not discretion to [*159] abandon the gatekeeping function. I think it worth adding that it is not discretion to perform the function inadequately. Rather, it is discretion to choose among reasonable means of excluding expertise that is fausse and science that is junky. Though, as the Court makes clear today, the Daubert factors are not holy writ, in a particular case the [***257] failure to apply one or another of them may be unreasonable, and hence an abuse of discretion.

DISSENTBY: STEVENS (In Part)

DISSENT:

JUSTICE STEVENS, concurring in part and dissenting in part.

The only question that we granted certiorari to decide is whether a trial judge "may . . . consider the four factors set out by this Court in Daubert v. Merrell Dow Pharmaceuticals, Inc., 509 U.S. 579, 125 L. Ed. 2d 469, 113 S. Ct. 2786 (1993), in a Rule 702 analysis of admissibility of an engineering expert's testimony." Pet. for Cert. i. That question is fully and correctly answered in Parts I and II of the Court's opinion, which I join.

Part III answers the quite different question whether the trial judge abused his discretion when he excluded the testimony of Dennis Carlson. Because a proper answer to that question requires a study of the record that can be performed more efficiently by the Court of Appeals than by the nine Members of this Court, I would remand the case to the Eleventh Circuit to perform that task. There are, of course, exceptions to most rules, but I firmly believe that it is neither fair to litigants nor good practice for this Court to reach out to decide questions not raised by the certiorari petition. See General Electric Co. v. Joiner, 522 U.S. 136, 150-151, 139 L. Ed. 2d 508, 118 S. Ct. 512 (1997) [**1180] (STEVENS, J., concurring in part and dissenting in part).

Accordingly, while I do not feel qualified to disagree with the well-reasoned factual analysis in Part III of the Court's opinion, I do not join that Part, and I respectfully dissent from the Court's disposition of the case.

REFERENCES:

31A Am Jur 2d, Expert and Opinion Evidence 342; 63B Am Jur 2d, Products Liability 1852, 1871, 1873

USCS Court Rules, Federal Rules of Evidence, Rule 702

L Ed Digest, Evidence 641, 643

L Ed Index, Expert and Opinion Evidence; Products Liability; Rules of Evidence; Tires and Wheels

Annotation References:

Reliability of scientific technique and its acceptance within scientific community as affecting admissibility, at federal trial, of expert testimony as to result of test or study based on such technique--modern cases. 105 ALR Fed 299.

When will expert testimony "assist trier of fact" so as to be admissible at federal trial under Rule 702 of Federal Rules of Evidence. 75 ALR Fed 461.

Products liability: admissibility of expert or opinion evidence that product is or is not defective, dangerous, or unreasonably dangerous. 4 ALR4th 651.

BIBLIOGRAPHY

Baer, Hans A. Toward a political ecology of health. *Medical Anthropology* 10(4): 451-454.

Basaglia, Franco. Breaking the Circuit of Control. *Critical Psychiatry* (D. Ingleby, ed.). New York: Pantheon, 1980.

Basaglia, Franco. *Scritti I: Dalla Psichiatria Femonenologica all' Esperienza di Gorizia.* Turin: Einaudi, 1981.

Bateson, Gregory. *Mind and Nature: A Necessary Unity.* New York: Dutton, 1979.

Bidney, David. 1953. *Theoretical Anthropology.* New York: Columbia University Press.

Black, Bert and David E. Libenfeld. Epidemiologic Proof in Toxic Tort Litigation. 52 Fordham L. Rev. 732 (1984).

Black, Bert. A Unified Theory of Scientific Evidence, 56 *Fordham Law Review.* 595 (1988).

Black, Bert. Subpoenas and Science – When Lawyers Force their Way into the Laboratory. 336 New Eng. J. Med. 725 (1997).

Bridgman, P.W. *The Logic of Modern Physics.* New York: Macmillan, 1927.

Broad, William and Nicholas Wade. *Betrayers of the Truth; Fraud and Deceit in the Halls of Science.* New York: Simon and Schuster, 1982.

Brody, Howard. *Placebos and the Philosophy of Medicine; Clinical, Conceptual and Ethical Issues.* Chicago: University of Chicago Press, 1980.

Brush, S.G. Should the History of Science be X-rated? *Science* 183:1164-72, 1974.

Burtt, E.A. *The Metaphysical Foundations of Modern Science* (revised ed.). New York: Garden City, 1954.

Cass (Coronary Artery Surgery Study) Principal Investigators and their Associates. A randomized trial of coronary artery bypass surgery: Survival Data. *Circulation* 63:939-9450, 1983.

Castiglioni, Arturo. *The Renaissance of Medicine in Europe,* 1934.

Cawte, John. *Medicine Is the Law; Studies in Psychiatric Anthropology of Australian Tribal societies.* Honolulu: University Press of Hawaii, 1974.

Chesebro, Kenneth J. Galileo's Retort: Peter Huber's Junk Scholarship. *The American University Law Review* Vol 42: 1637-1691, 1993.

Crozatier, B.J. Ross, D. Franklin, et al. Myocardial Infarction in the Baboon: Regional Function and the Collateral Circulation. *American Journal of Physiology: Heart Circulation Physiology* 235:H413-H421, 1978.

Daubert v. Merrill Dow Pharm., Inc. 727 F. Supp. 570, 571 (S.D. Cal 1989) [hereinafter Daubert I], aff'd 951 I. 2d 1128 (9th Circ. 1991) [hereinafter Daubert II] vacated and remanded, Daubert III, 509 v.s. 579, aff'd, Daubert IV, 43 F. 3d 1311 (1995).

Descartes, René. *Discourse on Method and the Meditations* (F.E. Sutcliffe, trans.). Middlesex: Penguin, 1968 (first published 1637. Leyden).

Devereux, George. *Basic Problems of Ethnopsychiatry.* Chicago: University of Chicago Press, 1980.

Devereux, George. *From Anxiety to Method in the Behavioral Sciences.* Paris and the Hague: École Pratique des Hautes Études and Mouton, 1967.

Eddy, D.M., and C.H. Clanton. The art of diagnosis; Solving the clinicopathological exercise. *New England Journal of Medicine* 21:1263-1268, 1982.

Edwards, Paul, ed. *The Encyclopedia of Philosophy,* p. 344, New York: Macmillan and Free Press, 1967.

Ell, Stephen R. Interhuman Transmission of Medieval Plague. *Bulletin of the History of Medicine* 54:497-510, 1980.

Elstein, A.S. *Medical Problem Solving: An Analysis of Clinical Reasoning.* Cambridge, MA: Harvard University Press, 1978.

Etkin, Nina L. (ed). *Plants in Indigenous Medicine and Diet: Biobehavioral Approaches.* Bedford Hills, NY: Redgrave, 1986.

Feigl, Herbert. The Mind-Body Problem in the Development of Logical Empiricism. *Readings in the Philosophy of Science* (H. Feigl and M. Brodbeck, eds) pp. 612-626. New York: Appleton-Century-Crofts, Inc., 1953.

Feyerabend, Paul, ed. *Against Method.* London: Verso, 1975.

Fleck, Ludwig. Genesis and Development of a Scientific Fact (Thaddeus J. Trenn and Robert K. Merton, eds.). Chicago: University of Chicago Press, 1979[1935].

Fortune, Reo. *Manus Religion.* Philadelphia: American Philosophical Society, 1935.

Foucault, Michel. *The Birth of the Clinic.* New York: Vintage, 1975.

Foucault, Michel. *The Archaeology of Knowledge and the Discourse on Language.* New York: Pantheon, 1969.

Frank, Philipp. *Philosophy of Science: the Link between Science and Philosophy.* Englewood Cliffs, N.J.: Prentice Hall, 1951.

Frye v. U.S. 293F. 1013 (D.C. Cir. 1923).

Gaudet, Carol A. Note: linking genes and behavior: the social and legal implications of using genetic evidence in criminal trials. 24 Fordham Urb L.J. 597 (1997).

Giannelli, Paul C. and Edward J. Imwinkelreid. *Scientific Evidence* (2nd ed 1993 and Supp 1994).

Gianni, A.J. The admissibility of scientific evidence in psychiatric malpractice: Junk science and the Daubert case. *Journal of Clinical Forensic Medicine* 1: 145: 1994.

Goodfield, June. *Cancer Under Siege.* London: Hutchinson, 1975.

Goodwin, J.S., J.M. Goodwin and A.V. Vogel. Knowledge and use of placebos by house officers and nurses. *Annals of Internal Medicine* 91:106-110, 1979.

Haack, Susan. *Evidence and Inquiry: towards Reconstruction in Epistemology.* Oxford, UK and Cambridge, USA. Blackwell, 1993.

Halvey, David N., ed. *The Merck Manual of Diagnosis and Therapy.* Rahway, NJ: Merck Sharp and Dohme, 1972.

Harvey, William. *Exercitatio Anatomica de Motu Cordis et Sanguinis in Animalibus.* (facsimile of 1628 Francofurti ed., Keynes English translation of 1928). Birmingham: Classics of Medicine Library, 1978.

Hite, C. Bridging the U.S.-Soviet Psychiatric Gap. *Psychiatric News,* 9 (part I):6-17; 19 (part 2):30-32, 40, 1974.

Huber, Peter. *Galileo's Revenge: Junk Science in the Courtroom.* New York Basic Books, 1995.

Jacobs, Thomas Jefferson. *Scenes, Incidents and Adventures in the Pacific Ocean.* New York: Harpen, 1844.

Jaffee, Leonard R. *Of Probativity and Probability: Statistics, Evidence and the Calculative Chance at trial.* 46 V. Pitt L. Rev. 925, 928, 929 (1985).

James, William. Pragmatism Lectures. Boston Lowell Institute. 1907. New York: 1987.

Janzen, John M. *The Quest for therapy in Lower Zaire.* Comparative Studies in Health Systems and Medical care. No. 1 Berkeley: California University Press, 1978.

Kahneman, D. (ed) *Judgment under Uncertainty: Heuristics and Biases,* New York: Cambridge University Press.

Kant, Immanuel. *The Critique of Pure Reason.* 2nd edition. (N. Kemp Smith, tr.). Chicago: University of Chicago Press, 1933.

Katz, Pearl. Ritual in the Operating Room. Ethnology 20:335-350, 1981.

Katz, Pearl. *The Scalpel's Edge; The Culture of Surgeons.* Boston: Allyn and Bacon, 1999.

Kleinman, Arthur. *Patients and Healers in the Context of Culture: An Exploration of the Borderland between Anthropology Medicine, and Psychiatry.* Berkeley: University of California Press, 1980.

Knox, Richard. The Harvard Fraud Case: Where Does the Problem Lie? *Journal of the American Medical Association* 249(14): 1797-1803, 1983.

Kuhn, Thomas S. *The Structure of Scientific Revolutions* Second ed. Chicago: University of Chicago Press, 1970.

Kuipers, Joel C. Medical Discourse in Anthropological Context; Views of Language and Power. *Medical anthropology Quarterly* 3:2: June 1989.

Kulynych, Jennifer. Psychiatric neuroimaging evidence: a high-tech crystal ball? 49 Stan L. Rev. 1249 (1997).

Lakatos, Imre. *Mathematics, Science and Epistemology.* 1978a.

Lakatos, Imre. *The Methodology of Scientific Research Programmes.* John Worall and Gregory Currie, (eds), 1978b.

Lanternari, Vittorio. *The Religion of the Oppressed: a Study of Messianic Cults.* New York: Knopf, 1963.

Latour, Bruno, and Steve Woolgar. *Laboratory Life; The Social Construction of Scientific Facts.* Sage Library of Social Research, vol. 80. London: Sage Publications, 1979.

Lee, K.L., F. McNeer, C.F. Starmet, et al. Clinical judgment and statistics; Lessons from a simulated randomized trial in coronary artery disease. *Circulation* 61:508-515, 1980.

Leiter, Brian. The Epistemology of Admissibility: Why even good Philosophy of Science would not make for good Philosophy of Evidence. 1997. B.Y.U.L. Rev. 803.

Leonard, Jaffee R. Of Probativity and Probability: Statistics, Evidence, and the Calceileis of Chance at Trial. 46 U. Pitt. L. Rev. 925, 928, 929 (1985).

Lock, Margaret and Deborah Gordon (eds). *Biomedicine Examined.* Dodrecht, Holland: Kluwer Academic Publishers, 1988.

Lonergan, Bernard. *Insight: Study of Human Understanding.* New York: Philosophical Library: Longmans, Green, 1965.

Longhurst, John C., and John Ross, Jr. Extracardiac and Coronary Vascular Effects of Digitalis. Symposium on William Withering and the Foxglove: the 200th Anniversary of His First Report. *Journal of the American College of Cardiology* 5(5):99A-105A, 1985.

Longino, Helen E. *Science as Social Knowledge: Values and Objectivity in Scientific Inquiry*, 1990.

McElroy, Ann and Patricia K Townsend (eds). *Medical Anthropology in Ecological Perspective.* 3rd edition. Boulder, Co: Westview Press, 1996.

Mead, Margaret. *New Lives for Old: Cultural Transformation-Manus.* 1928-1953. New York: Morrow, 1956.

Medical Letter on Drugs and Therapeutics. 21(2), 1979.

Medical Letter on Drugs and Therapeutics. 23(3):9, 1981.

Melton, Gary B. (ed). *The Law as a Behavioral Instrument.* Lincoln: University of Nebraska press, 1986.

Miller, George A. *Language and Communication.* New York: McGraw Hill, 1951.

Mills, John. Stuart, *Systems of Logic.* London, 1849.

Mishima, Yukio. *Runaway Horses.* New York: Vintage International, Division of Random House. 1990, p. 365. Tr. from the Japanese by Edward G. Seindensticker.

Moerman, Daniel E. General medical effectiveness and human biology: placebo effects in the treatment of ulcer disease. *Medical Anthropology Quarterly* 14(3):14-16, 1983.

Moerman, Daniel E. *Native American Ethnobotany.* Portland, Oregon: Timker Press, 1993.

Moore, Michael. *Law and Psychiatry: Rethinking the Relationship.* Cambridge University Press, 1984.

Morgagni, Giovanni Battista. *De Sedibus et Causus Morborundum per Anatomen Idagatis* (often reprinted. First English edition, 1769).

Nader, Laura. *Harmony Ideology: Justice and Control in a Zapotec Mountain Village.* Stanford, California: Stanford University Press, 1990.

Nagel, Ernest. *The Structure of Science; Problems in the Logic of Scientific Explanation.* N.Y. Harcourt and Brace 1961.

Neverman, Hans. *Admiralitäts-Inseln. II. Ethnographic: A Melanesien. Erqebnisse der Südsee-Expedition, 1908-1910.* G. Thelanius, Hamburg, Friederichsen, De Gruyter, Vol. 3. 1934.

Pierce, Charles Sanders. *Philosophical writings of Pierce.* (Justus Buchler, ed.) 1955.

Popper, Karl R. *The Myth of Framework* 8 (M.A. Notturno, ed.) 1994.

Popper, Karl R. *Objective Knowledge.* Oxford: Oxford University Press, 1979, 2nd edition.

In re Quinlan. 1976. 70 N. J. 10, 355 A. 2d 679, Cert Denied 429 U.S. 922.

Radine, Lawrence B. Pangolins and advocates: Vulnerability and self-protection in a mental patients' rights agency. *The Anthropology of Medicine* (L. Romanucci-Ross, D. Moerman, and L. Tancredi, eds.), pp. 366-387. South Hadley, MA: Bergin, 1983.

Reich, Walter. *Diagnostic Ethics: The Uses and Limits of Psychiatric Explanation* (S. Block and P. Chodoff, eds.), pp. 61-88. Oxford: Oxford University Press, 1981.

Reichenbach, Hans. *The Rise of Scientific Philosophy.* Berkeley. Univ. of California Press, 1951.

Reinmann, Bernard. *On the Hypothesis upon which Geometry is Founded*, 1867.

Rollins, N. *Child Psychiatry in the Soviet Union.* Cambridge, MA: Harvard University Press, 1972.

Romanucci-Ross, Lola. Creativity in illness: Methodological Linkages to the Logic and Language of Science in Folk Pursuit of Health in Central Italy. *Social Science and Medicine.* 23: 1: 1-7, 1986.

Romanucci-Ross, Lola. *Mead's Other Manus: Phenomenology of the Encounter.* Westport, Conn. Bergin and Garvey, 1983.

Romanucci-Ross, Lola. The Impassionate Cogito: Shaman and Anthropologist. *Shamanism; Past and Present.* Mihály Hoppál and Otto von Sadovszky. Ethnographic Institute, Hungarian Academy of Sciences and Los Angeles: Istor Books, 1989, pp. 35-43.

Romanucci-Ross, Lola, Daniel E. Moerman and Laurence R. Tancredi (eds). *The Anthropology of Medicine; from Culture to Method.* Westport, Conn. and London: Bergin and Garvey. 3rd edition. 1997.

Romanucci-Ross, Lola and Laurence R. Tancredi. Psychiatry, the Law and Cultural Determinants of Behavior. *International Journal of Law and Psychiatry.* 9:265-193. New York: Pergamon Press, 1986.

Rougier, Louis. *La Philosophie Geometrique de Henri Poincare.* Paris: F. Alcan, 1920.

Sanders, M., F. White, and C. Bloor. Cardiovascular responses of dogs and pigs exposed to similar physiological stress. *Comparative Physiology and Biochemistry* 58(A):365-370.

Scheper-Hughes, Nancy. *Death without Weeping: the Violence of Everyday Life in Brazil.* Berkeley: University of California Press, 1992.

Schlick, M. *Allgemeine Erkenntnislehre.* 2nd edition. Berlin: Springer, 1925.

Schuck, Peter H. Multi-culturalism redux: science, law and politics. II Yale Law and Policy Review I, 15. 1993

Schwartz, Lola R. (aka Lola Romanucci-Ross) The Hierarchy of Resort in Curative Practices: The Admiralty Islands. Melanesia. 10:3:201-209. Sept. 1969.

Schwartz, Theodore. *The Paliau Movement in the Admiralty-Islands,* 1946-54. Anthropological Papers of the American Museum of Natural History. Part 2, 207-422. New York. 1962

Schwartz, Theodore. Cultural Totemism: Ethnic Identity, Primitive and Modern. In: *Ethnic Identity, Creation, Conflict and Accommodation.* Lola Romanucci-Ross and George De Vos, Eds. Walnut Creek, California and London: Altamira Press. 48-72. 1995.

Selzer, Richard. *Mortal Lessons; Notes on the Art of Surgery.* New York: Simon and Schuster, 1976.

Shapiro, A.K. The placebo response, *Modern Perspectives in World Psychiatry* (J.G. Howells, ed.). Edinburgh: Oliver and Boyd.

Sinha, Arushi. An Overview of Telemedicine: The Virtual Gaze of Health Care in the Next Century. *Medical Anthropology Quarterly* 14: 30: 291-369. Sept. 2000.

Snellen, H.A. *History of Cardiology.* Rotterdam: Donker Academic Publications, 1984.

Snyder, J.W. Inquiry and causation on trial; the phenomenon of multiple chemical sensitivities: 2 Wid. L. Symp J. 97 (1997).

Spiegelberg, H. *The Phemonological Movement.* The Hague: Mouton, 1960.

Tancredi, Laurence. Psychiatry and social control. *The Anthropology of Medicine* (L. Romanucci-Ross, D. Moerman, and L. Tancredi, eds.), pp. 284-297. South Hadley, MA: Bergin, 1960.

Tancredi, Laurence R. and David Weisstub. Ideology and Power: Epidemiology and interpretation in law and psychiatry. L. Romanucci-Ross; D. Moerman and L. Tancredi. *The Anthropology of Medicine* 2nd Edition, 1991.

Tomoike, H., D. Franklin, S.W. McKown, et al. Functional evaluation of coronary collateral development in conscious dogs. *American Journal of Physiology: Heart Circulation Physiology* 241:H519-H524, 1981.

Trostle, James A. *Medical Compliance as an Ideology, Social Science and Medicine* vol. 27, no. 12, pp. 1299-1308, 1988.

Wilder, Raymond. *Introduction to the Foundation of Mathematics.* pp. 256-261. New York: Wiley, 1952.

Williams, Bernard. René Descartes. *The Encyclopedia of Philosophy,* vol. 1, pp. 344-354, 1967.

Woolf, Patricia K. Deception on Scientific Research. In *American Association for the Advancement of Science, Project on Scientific Fraud and Misconduct*: Report of Workshop Number One 37, 78. 1988.

World Health Organization. *Report of the International Pilot Study of Schizophrenia,* vol. 1. Geneva: WHO, 1973.

Worsely, Peter. *The Trumpet Shall Sound: a Study of Cargo Cults in Melanesia.* London: MacGibbon and Kee, 1957.

Wright, P. and A. Treacher. *The Problem of Medical Knowledge; Examining the Social Construction of Medicine.* Edinburgh: Edinburgh University Press, 1982.

INDEX

International Library of Ethics, Law, and the New Medicine

1. L. Nordenfelt: *Action, Ability and Health.* Essays in the Philosophy of Action and Welfare. 2000 ISBN 0-7923-6206-3
2. J. Bergsma and D.C. Thomasma: *Autonomy and Clinical Medicine.* Renewing the Health Professional Relation with the Patient. 2000 ISBN 0-7923-6207-1
3. S. Rinken: *The AIDS Crisis and the Modern Self.* Biographical Self-Construction in the Awareness of Finitude. 2000 ISBN 0-7923-6371-X
4. M. Verweij: *Preventive Medicine Between Obligation and Aspiration.* 2000 ISBN 0-7923-6691-3
5. F. Svenaeus: *The Hermeneutics of Medicine and the Phenomenology of Health.* Steps Towards a Philosophy of Medical Practice. 2001 ISBN 0-7923-6757-X
6. D.M. Vukadinovich and S.L. Krinsky: *Ethics and Law in Modern Medicine.* Hypothetical Case Studies. 2001 ISBN 1-4020-0088-X
7. D.C. Thomasma, D.N. Weisstub and C. Hervé (eds.): *Personhood and Health Care.* 2001 ISBN 1-4020-0098-7
8. H. ten Have and B. Gordijn (eds.): *Bioethics in a European Perspective.* 2001 ISBN 1-4020-0126-6
9. P.-A. Tengland: *Mental Health.* A Philosophical Analysis. 2001 ISBN 1-4020-0179-7
10. D.N. Weisstub, D.C. Thomasma, S. Gauthier and G.F. Tomossy (eds.) : *Aging: Culture, Health, and Social Change.* 2001 ISBN 1-4020-0180-0
11. D.N. Weisstub, D.C. Thomasma, S. Gauthier and G.F. Tomossy (eds.) : *Aging: Caring for our Elders.* 2001 ISBN 1-4020-0181-9
12. D.N. Weisstub, D.C. Thomasma, S. Gauthier and G.F. Tomossy (eds.) : *Aging: Decisions at the End of Life.* 2001 ISBN 1-4020-0182-7
    (Set ISBN for Vols. 10-12: 1-4020-0183-5)
13. M.J. Commers: *Determinants of Health: Theory, Understanding, Portrayal, Policy.* 2002 ISBN 1-4020-0809-0
14. I.N. Olver: *Is Death Ever Preferable to Life?* 2002 ISBN 1-4020-1029-X
15. C. Kopp: *The New Era of AIDS.* HIV and Medicine in Times of Transition. 2003 ISBN 1-4020-1048-6
16. R.L. Sturman: *Six Lives in Jerusalem.* End-of-Life Decisions in Jerusalem - Cultural, Medical, Ethical and Legal Considerations. 2003 ISBN 1-4020-1725-1
17. D.C. Wertz and J.C. Fletcher: *Genetics and Ethics in Global Perspective.* 2004 ISBN 1-4020-1768-5
18. J.B.R. Gaie: *The Ethics of Medical Involvement in Capital Punishment.* A Philosophical Discussion. 2004 ISBN 1-4020-1764-2
19. M. Boylan (ed.): *Public Health Policy and Ethics.* 2004 ISBN 1-4020-1762-6; Pb 1-4020-1763-4
20. R. Cohen-Almagor: *Euthanasia in the Netherlands.* The Policy and Practice of Mercy Killing. 2004 ISBN 1-4020-2250-6
21. D.C. Thomasma and D.N. Weisstub (eds.): *The Variables of Moral Capacity.* 2004 ISBN 1-4020-2551-3
22. *To be published.*
23. P. McCullagh: *Conscious in a Vegetative State? A Critique of the PVS Concept.* 2004 ISBN 1-4020-2629-3

International Library of Ethics, Law, and the New Medicine

24. L. Romanucci-Ross and L.R. Tancredi: *When Law and Medicine Meet: A Cultural View*. 2004
ISBN 1-4020-2756-7

KLUWER ACADEMIC PUBLISHERS – DORDRECHT / BOSTON / LONDON

Lightning Source UK Ltd.
Milton Keynes UK
15 December 2009

147562UK00001B/43/A

9 781402 027567